Humans
and
Automation

Humans
and
Automation

System Design and Research Issues

Thomas B. Sheridan

A John Wiley & Sons, Inc., Publication
Wiley Series in System Engineering and Management

HFES Issues in Human Factors and Ergonomics Series, Volume 3
Supervising Editor: David Meister

Published in cooperation with the Human Factors and Ergonomics Society
P.O. Box 1369 Santa Monica, CA 90406-1369 USA
310/394-1811 Fax 310/394-2410 info@hfes.org http://hfes.org

A JOHN WILEY & SONS, INC., PUBLICATION

for James, Jake, Nicholas, Nora, Spencer, Charlie, Anna, Sam, and Vanessa,
with hope that automation will enrich their lives and not the opposite

Individual readers of this book and nonprofit libraries acting for them are freely permitted to make fair use of the material in it, such as to copy a chapter for use in teaching or research. Permission is granted to quote excerpts from chapters in scientific works with the customary acknowledgment of the source, including the author's name, the book's title, and the publisher's name.

Permission to reproduce any chapter or a substantial portion (more than 300 words) thereof, or any figure or table, must come from the author and from the HFES Communications Department. Reproduction or systematic or multiple reproduction of any material in this book is permitted only under license from the Human Factors and Ergonomics Society. Address inquiries to the Communications Department, Human Factors and Ergonomics Society, P.O. Box 1369, Santa Monica, CA 90406-1369 USA; 310/394-1811, fax 310/394-2410, e-mail lois@hfes.org.

Additional copies of this book may be obtained from the Human Factors and Ergonomics Society. Discounts apply on purchases of five copies or more; contact the Society at the address given above.

The Human Factors and Ergonomics Society is a multidisciplinary professional association of 5000 persons in the United States and throughout the world. Its members include psychologists, designers, and scientists, all of whom have a common interest in designing systems and equipment to be safe and effective for the people who operate and maintain them.

Library of Congress Cataloging-in-Publication Data is available.

ISBN 0-471-23428-1

10 9 8 7 6 5 4 3 2 1

CONTENTS

Preface   x

PART I: BACKGROUND

Chapter 1. Introduction   2
     Human Factors: What Is It (Are They)?   2
     What Is Automation?   9
     A Preview of the Critical Issues   12
       of Humans and Automation

Chapter 2. Human Interaction with   14
     Automation in Various Contexts
     Aircraft and Air Traffic Control   14
     Automobiles and Highway Systems   23
     Trains   27
     Ships   31
     Spacecraft   32
     Teleoperators for Hazardous   33
       Environments, Motion Rescaling,
       and Other Nonroutine Tasks
     Virtual Reality for Training   36
       and Entertainment
     Nuclear Power Plants and   39
       Process Control
     Manufacturing of Discrete Products   41
     Hospital Systems   43
     Command and Control in Defense   46
       and Civilian Emergencies
     Office Systems   47
     Education   48
     The "Smart" Home   49
     Automation on the Person   51

PART II: DESIGN OF HUMAN-AUTOMATION SYSTEMS

Chapter 3. The Analysis and Design Process   **54**

The Role of the Human Factors   54
Professional

Human-Machine Task Analysis: 55
Goals and Constraints

Human-Machine Function Allocation   57

Large System Organization   65

Optimization   66

Handbooks, Guidelines, and Standards   67

Chapter 4. Human Performance in   **69**
Relation to Automation

Nominal Performance of Humans: 69
Speed and Accuracy

Robustness, Adaptivity, and Self-   71
Pacing in Continuous Tasks

Biases of Human Decision Makers   72

Situation Awareness   75

Trust   76

Mental Workload   77

Human Error   82

Human-Machine System   89
Reliability Analysis

Chapter 5. Displays and Decision Aids   **95**

Display Dedication, Integration,   95
and Flexibility

Ecological Displays, Compatibility,   99
Animation, and Virtual Reality

Interaction through Icon and Menu   103
  Displays, Voice, and Haptic Inputs
Alarms and Warnings, and   104
  the Problem of Monitoring
  for Abnormalities
Displays to Aid with Monitoring and   107
  Procedures
Model-Based Predictor Displays   108
  to Aid with Control
Decision Aids for Satisficing   110
Etiquette of Conversing with a   112
  Decision Aid
The Ultimate Decision Aid   113

Chapter 6. Supervisory Control   115
Definition of Supervisory Control   115
The Historical Trend   116
Five Roles of the Human Supervisor   118
When to Automate   128
  (and Use Supervisory Control)

Chapter 7. Simulation, Testing, and Evaluation   130
Human-Automation Design as   130
  an Empirical Activity
Human-in-the-Loop Simulation   131
Estimation Theory as a Metaphor   135
  for Using Simulation in the Research
  and Design Process
Caveats on Designing Experimental   136
  Tests
Virtual Engineering   138
Evaluation   139

PART III: GENERIC RESEARCH ISSUES

Chapter 8. Technical Issues of Humans   **144**
and Automation

System Complexity: How to Cope   144
Monitoring: A Job Nobody Wants   147
Giving and Taking Advice   148
How Far to Go with Automation   150
Ultimate Authority: When and   152
   by Whom
Human-Centered Automation: 155
   What Does It Mean?
Limits of Modeling and Prediction   157
   in Human-Automation Systems
Limits on Design of Human-   160
   Automation Systems

Chapter 9. Social Issues of Humans and   **163**
Automation

Social Benefits of Automation   163
Human Irrationality   165
Alienation of the Individual   170
Alienation of the Community   175

Chapter 10. System Management and   **182**
Education

System Management   182
Education   186

Appendix A. Elements of the Most Important   **189**
Normative Models

     1. Probability and Some Probabilistic   189
        Tests
     2. Information   194
     3. Utility Function (Objective Function)   200
     4. Discrete Decisions   203
     5. Continuous Control   220

Appendix B. Sample Problems and   **231**
Solutions on Normative Models

     1. Problems on Topics Covered in   231
        Appendix A
     2. Solutions to Problems Stated in   237
        Appendix B

References   **245**

Author Index   **257**

Subject Index   **261**

PREFACE

This book is meant as a succinct tutorial for professionals and students who want to understand the many human interactive aspects of automation. It is designed to serve engineers who know a bit about automation and want to learn what the field of human factors is and how it relates to automation. It is also designed for human factors specialists, behavioral and life scientists, and physicians who want to learn how humans interact with automation.

The book is intended as a codification and review of what is known — what is reasonably well established and consensual regarding humans and automation, including relevant mathematical models. It is not intended as a presentation of the latest research in the field or in specialized domains of it, though there are ample references to the research sources. Other books — for example, Parasuraman and Mouloua (1996), Mouloua and Koonce (1997), and Sarter and Amalberti (2000) — are proceedings of recent meetings or have chapters written by specialists on their own recent research.

The book was written at the invitation of the Human Factors and Ergonomics Society. I hope it can be useful for emerging academic courses in human factors and automation as well as serving as a general reference. I have tried to write in plain English, avoiding jargon where possible but nevertheless introducing readers to technical terms, common acronyms, and associated concepts with which they should become familiar.

The text of this book was essentially complete prior to the horrific events of September 11, 2001. However, in retrospect, it is clear that the problems of humans and automation have become more important than ever. Indeed, we need improved automation for our security and well-being as a society, yet we realize how vulnerable are our various large-scale aggregates of humans and automatic equipment (transportation, communication, buildings, supply of food, water, electricity, etc.) to terrorist acts. The challenge is immense.

Organization of the Book

The book is organized into 10 chapters in three parts. Part I, "Background," contains Chapter 1, the introduction; and Chapter 2, a sampling of automation that has been implemented in various system contexts.

Chapter 1 defines human factors engineering, gives some history, and discusses the methodology that has grown up around this field. It also defines automation and gives a quick summary of the basic ideas of classical feedback control essential for understanding automatic control. Readers who are familiar with either human factors/ergonomics or automation may skip these introductory explanations, or they may use them to check whether they share my assumptions.

Chapter 2, "Human Interaction with Automation in Various Contexts," discusses the nature of current and predicted future interactions between humans and automation in areas that previously were controlled by humans only. The contexts include aviation (both piloting and air traffic control), automobiles and highway systems, trains, ships, spacecraft, robotic teleoperation in hazardous environments, virtual reality for training and entertainment, nuclear power plants and process control, discrete parts manufacturing, hospital systems, command and control in defense and civilian emergencies, office systems, education, automation in the "smart home," and automation on the person. The reader will begin to observe that the human factors problems of automation may differ in detail but really are fairly common to all of these contexts.

Part II, "Design of Human-Automation Systems," is the longest part. This theme is detailed in Chapters 3 through 7. Chapter 3, "The Analysis and Design Process," reviews task analysis, allocation of functions between humans and machines, and some ideas about modeling and optimization.

Chapter 4, "Human Performance in Relation to Automation," discusses speed and accuracy, robustness, adaptivity, self-pacing, situation awareness, mental workload, human error, system reliability, and decision making.

Chapter 5, "Displays and Decision Aids," is not intended to be a tutorial on the standard human factors of display design. Rather, it includes commentary on display dedication, integration, and flexibility; ecological displays; means of interacting with displays; alarms and warnings; display aids for monitoring and procedures; predictor displays; decision aids for satisficing; and decision-aiding etiquette.

Chapter 6, "Supervisory Control," defines this broad concept, which is really the human counterpart of automation. Sections on history, human roles, and when to automate and use supervisory control are included.

Chapter 7, "Simulation, Testing, and Evaluation," discusses the empirical nature of human-automation design, human-in-the-loop simulation, estimation theory as a metaphor for simulation in design, caveats on designing experimental tests, virtual engineering, and the evaluation process.

Part III, "Generic Research Issues," is divided into Chapters 8, 9, and 10. Chapter 8, "Technical Issues of Humans and Automation," examines the variety of problems that researchers attempt to solve through scientific means. These include complexity, monitoring, giving and taking advice, how far to go with automation, ultimate authority, human-centered automation, limits of modeling and prediction in human-automation systems, and limits of design.

In Chapter 9 I discuss the broader social and philosophical issues that tend to swirl around ever-increasing automation relative to humans as individuals and as a society. Included are discussions on automation benefits, human irrationality, alienation of the individual, alienation of the community, and educational needs.

Chapter 10 very briefly emphasizes some issues that must be dealt with to ensure than human-automation system projects are properly managed so as to come to fruition and be acceptable in the context of the larger social milieu. It also provides caveats regarding educational needs for the designers, managers, and the public.

Following Chapter 10 is an extensive appendix (A), which provides a tutorial on the normative theories that are believed to be most important for dealing with research or design on human-automation interaction. Sections of Appendix A include probability and some probabilistic tests (Bayes, sampling, null hypothesis tests), theories of information and information value, utility, discrete decisions (Pareto frontier, crisp and fuzzy rule-based decisions, decisions under uncertainty, decisions of whether to automate, game theory, signal detection theory, fuzzy signal detection, decision maker calibration), and continuous control (on-off control, transfer functions, stability, gain-phase plots, control optimization, fuzzy control).

Appendix B provides some sample problems and solutions for models presented in Appendix A.

Acknowledgments

I gratefully acknowledge the contributions of the following persons: Russ Ferrell for collaborating with me on an earlier book, which seeded much of Appendix A of this book; John Hansman, Joachim Meyer, Bill Rouse, Andy Sage, and Toshi Inagaki for commenting on various chapters; Neville Moray for many fruitful discussions on the issues; numerous students who were also my teachers; and of course Rachel for being Rachel.

PART I: BACKGROUND

Chapter 1

INTRODUCTION

This section presents some basic ideas of human factors (as the term is used here; see below) and automation, as separate and distinct entities and emerging disciplines. Because this is meant to be a primer, the intention is not to give a complete and comprehensive treatise on either topic. Many books have been written about each, especially about automation. But I want to start with the separate entities and then illustrate the coming together: the interactions, the cooperations (and lacks thereof), the marriage (however uncomfortable) between humans and automation. Does it make sense to talk about humans and automation interacting? Is a *human-automation system* an impossibility by definition, an oxymoron? The answer is a loud and clear "NO," and I will employ the term *human-automation system* from time to time without apology. Such a system is more restrictive than *human-machine system* in that it includes only those machines that — by a definition yet to be made — are automated. But they are not fully automated, not 100%. Actually, few systems are 100% auto-mated, with no human interaction whatsoever. Many, many systems are automated to a large extent and *also* have critical human interactions.

In the case of automation per se, some small bit of mathematics is usually deemed essential to say much of anything. However, the purpose of this book is to convey the basic concepts at an intuitive level. Therefore Appendix A presents the mathematical models fundamental to automation, whereas the main body of the book discusses the human-automation rela-tion qualitatively.

Human Factors: What Is It (Are They)?

Definitions
Human factors, also known as *human engineering* or *human factors engineering*, is the application of behavioral and biological sciences to the design of machines and human-machine systems. The most salient behav-ioral sciences are cognitive psychology and the broader field of experi-mental psychology (study of sensation, perception, memory, and thinking, as well as motor skill) and to a lesser extent organizational and social

psychology, sociology, and psychometrics, but not clinical psychology. The most salient biological science is human systems physiology (study of organ functions above the cellular level).

Often the term *ergonomics* (derived from the Greek *ergon*, work, and *nomos*, principle or law; literally, "work study") is used as a synonym for *human factors*. However, in some circles ergonomics has come to be associated with the narrower aspect of human factors dealing with anthropometry, biomechanics, and body kinematics as applied to the design of seating and workspaces. The interaction of humans with automation deals less with ergonomics in this traditional sense and more with sensory and cognitive function. The terms *cognitive engineering* and *cognitive ergonomics* have therefore come into vogue.

A Brief History

Throughout history there has been interest in the human body and brain and a practical interest in human performance in the workplace. Leonardo da Vinci made ingenious drawings of human-machine systems. In 1857 a Polish scholar, Wojciech Jastrobowski, published *An Outline of Ergonomics, or The Science of Work*. In the early 1900s Frederick Lewis Taylor, sometimes referred to as the "father of scientific management," tried to put the study of production line efficiency on a scientific basis using time and motion analysis (only to become a symbol of human exploitation several decades later by worker advocates). Edwin Link built early flight simulators at the time of World War I.

Human factors may be said to have begun in earnest as a discipline during World War II, when it was fully appreciated that modern weapons of war required explicit engineering of the interface between human and machine. This meant fitting the user to the machine (aircraft cockpits, radar or sonar workstations, gun sights, etc.) not only in terms of body size and strength but also with respect to limits of vision and hearing, experience, and learning capability. Psychologists, physiologists, and medical doctors in both the United States and Europe (primarily Britain and Germany) were brought into government laboratories to work with engineers and produce explicit system designs and do laboratory experiments to develop guidelines for future systems.

After the war most of these people returned to their domestic jobs (universities, companies, etc.) and brought with them an appreciation of the need to develop human factors as an engineering discipline. Their governments also appreciated the need and continued to support research in various aspects of human factors. In the United States the initial effort was mostly applied research on how to design displays and controls and workstations for defense systems, whereas in Europe the applications were mostly in industrial production, to help rebuild the damaged economies. Tools of human activity analysis from earlier production-line applications were refined and applied to a broad range of human workplaces. Measures of visual and auditory acuity and other aspects of human performance within work settings became commonplace. Several handbooks emerged from this period, notable among them the Van Cott and Kinkade *Human Engineering Guide to Equipment Design* (1972). This early, empirical phase has often been called (disparagingly) "knobs and dials engineering." The design of displays, controls, and workplace layouts, however, has remained as important as it ever was. The problems have just grown more complex, as the reader shall see.

During World War II a number of important theoretical concepts were developed, motivated by the need to build more capabilities into military systems. These included feedback control theory, signal detection theory, and decision theory. Shortly after that war information theory was formally introduced by Bell Laboratories. Initially these ideas were applied to physical systems with little thought to applying them to humans, but human factors engineers were quick to appreciate their broader applicability. The physical inputs and outputs became stimulus to human (what was presented on display devices) and response from human (what was measured from the control devices). The system engineering models became ways of characterizing human experimental data in laboratory experiments and field tests with displays and controls.

Perhaps more important, human factors engineers began to look at information, control, and decision making as a continuous process within a closed loop that also included physical subsystems — more than just sets of independent stimulus-response relations. (An appreciation for this difference may be gained by understanding the discussion of loop equations in Appendix A.) This stage of the development of human factors engineering brought it fully into the realm of systems engineering. Human input-output or transfer equations could be combined with input-output equations of the machinery. In fact, prior to this stage it made little sense to analyze the machine components of systems if the human component, connected in series, could not itself be analyzed in similar terms. So this might be called the "borrowed engineering models" stage of development of the human factors field.

In the early 1970s the computer was becoming more powerful, smaller, and cheaper, first as a minicomputer and a decade later as a microcomputer. This piece of technology changed human-machine systems profoundly. Computers especially changed display technology. Earlier displays were costly mechanical instruments that mostly had to be dedicated to one function. Now they are generated by computers with an infinite variety of display formats and animated graphics in living color. Computers also changed controls so that the human operator could type commands, use touch screens, or even speak commands and have them understood, in addition to using dedicated buttons, levers, wheels, or pedals. Also, as shall be seen, computers intervened in human-machine systems in many other ways, becoming intelligent agents to search data, give advice, and perform subordinate control functions. The problems for humans interacting with automation have become those of humans interacting with computers, but not only computers on desktops — more likely, computers in a thousand different systems in all walks of life.

Looking back, the progress in human factors or human-machine systems engineering might be summarized by three phases, as represented in Figure 1.1. Phase A (knobs and dials) refers to the traditional biomechanical ergonomics of workstation design. Phase B (borrowed engineering models) refers to the use of quantitative performance models of information processing, signal detection, and control that emerged after World War II. Phase C (human-computer interaction) refers to the variety of human interactions with computers that are occurring today. The development within each component over time is represented by a plot of "progress" (choose your own measure) versus time. Phases A and B have shapes depicting rapid growth at first, which then tapered off but have continued development. Phase C is still developing rapidly and constitutes most of the issues discussed in this book.

During the last half century, government agencies and companies have become more astute about the application of human factors. The U.S. Army, Navy, and Air Force each has a basic research arm with an active human factors program, and applied research is carried out in myriad

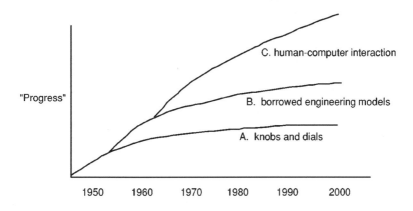

Figure 1.1 The history of intellectual development in human-machine systems engineering.

places, including the National Aeronautics and Space Administration (NASA), Federal Aviation Administration (FAA), Nuclear Regulatory Commission, and Department of Transportation. Professional societies have been founded, including the Human Factors and Ergonomics Society (HFES), which began in 1957 and now has about 5000 members, and the International Ergonomics Association, an umbrella organization that was organized at about the same time, now has more than 25 affiliated human factors/ergonomics societies around the world. The Institute of Electrical and Electronics Engineers (IEEE) Systems, Man, and Cybernetics Society has around 5000 individuals. Academic institutions have set up degree programs in human factors, mostly in psychology and industrial engineering departments; HFES publishes a listing of such programs. In Canada, Europe, and Japan a similar organizing of human factors activities has occurred.

Methods of Human Factors

In systems design, human factors takes the viewpoint of the human user. This is often overlooked by design engineers, who assume that because they themselves are human, the user's viewpoint and constraints will naturally be considered. The fallacy, of course, is that one person doesn't think like another person; a wide range of people must be accommodated. In particular, the designer, who may know the physical problems and physical process very well, does not think like the users (at least, not like 95% of them). So although human factors is often regarded as common sense, it is mostly very uncommon sense!

In contrast to the usual idea that *engineering design* means to design things or physical relations among things (call that *systems design*), *human factors design* means to design the relations between people and things. It requires a different perspective.

Do human factors theories exist? Not a lot, at least not in the same quantitative sense as in physics and engineering. Yes, there is theory in the qualitative sense — lots of that — and it is certainly not my intention to demean qualitative theory. (In fact, much of what will be said in this book can be called qualitative theory!) The quantitative theory that is available is limited to two types. First are those relatively few mathematical models that generalize the transducer properties of the eyes, ears, other sensors, and the musculoskeletal system. Then there are the information, decision, and control models, already referred to, that are borrowed from systems engineering. These are reviewed in Appendix A.

Human factors engineering is mostly an empirical activity, much like medicine, with a growing literature that must be interpreted in a practical context. The human factors practitioner almost always begins with some kind of task analysis to define what variables in what ranges are important (what kind of people to consider, how well trained they are, what motivations they have, etc.). Some part of the literature is always relevant, and, like the physician, the human factors practitioner should know how to access and make use of the experimental literature. Taking recommendations from handbooks and guidebooks may be worth doing for checking one's considerations, but it is definitely not recommended as the sole accession of the literature. Naive application of handbook data can lead to disaster!

It is usually essential to do some kind of ad hoc experiment to understand the problem and get ideas for its solution. Finally, after design solutions become tentative, it is often worth setting up some kind of computer simulation or mockup of the workplace and its displays and controls in order to train some real people who come close to the target population of users and to try out the prototype system with some performance scoring, error observations, and subjective assessments.

Good introductions to human factors engineering are Wickens, Gordon, and Liu (1998); Booher (1990); and Sanders and McCormick (1987).

The Shape of Human Factors Today

In Figure 1.1, Phase C in particular is growing very fast and in many ways, enhanced by artificially intelligent agents, human supervisory control, teleoperation, virtual reality, and many other new developments. Phase C, human-computer interaction, is synonymous with "humans and automation" in the broad sense (and see the definitions of automation in the next section).

Cognitive engineering and *cognitive ergonomics*, terms that were at first offensive to many because "cognitive" smacked of scientific imprecision, have become accepted as the challenge of dealing with human factors in automation. The point is that humans are doing less physical work than previously. Although the problems of humans doing physical work remain important, they are gradually being replaced by the problems of humans interacting mentally with computers and automation. Such interaction still requires seeing, hearing, and button pushing, but the human factors literature is in relatively good shape there. The problems are in attending, perceiving, remembering, deciding, and, in general, thinking — all aspects of that unfortunately still qualitative term, *cognition.*

One form of system representation is cited in this section, one that is both commonly accepted by cognitive engineers and highly relevant to humans and automation. This is Rasmussen's (1976, 1986) idea that human behavior is based on *skills, rules, and knowledge,* elements that are interdependent in the organism, as shown in Figure 1.2.

The interesting thing is that these elements have their counterparts in automation technology. Skill-based behavior corresponds to continuous closed-loop control. Rule-based behavior corresponds to rule-based computer decision making. Knowledge-based behavior, some say, has no counterpart in technology, but others assert that very advanced artificial intelligence comes close. The diagram in Figure 1.2 is an example of a cognitive engineering idea, one that seems to tie together human cognition and computer decision making/control.

Although human interaction with technology is recognized by the public as a salient issue and human factors as a professional discipline is now recognized in some engineering communities (e.g., aerospace and the military), the human factors profession is still struggling for recognition within much of the engineering world. This is particularly true of cognitive engineering or cognitive ergonomics. This is because engineers, having thoroughly utilized and become indoctrinated in Newtonian physics, expect all engineering to be up to that level of analytic codification. Engineers do appreciate that medicine is still an empirical science and that the human body in the health context cannot be expected to meet the standard of garden-variety physics. Somehow the point is missed that in engineering products and systems for people, there is a legitimate cognitive aspect that still must be treated empirically. This can be a frightening idea for engineers.

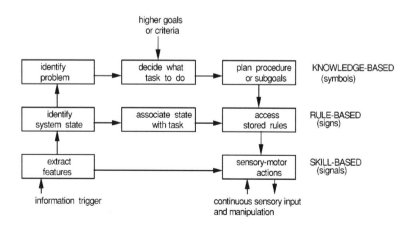

Figure 1.2 The Rasmussen knowledge-rule-skill hierarchy.

What Is Automation?

This section is not meant as a rigorous treatment of automation; rather, it provides an intuitive idea of the essential concepts. Familiarity with elementary notions of functions and differential equations is helpful; further exposition of quantitative ideas is in Appendix A. Readers can also find more rigorous treatments in the library. Those with a background in the theory of decision and control can skip what is already familiar.

Definitions

The term *automation* has several definitions. The *Oxford English Dictionary* (second edition, 1989) defines it as follows:

1. Automatic control of the manufacture of a product through a number of successive stages;
2. the application of automatic control to any branch of industry or science;
3. by extension, the use of electronic or mechanical devices to replace human labor

The term's first use is traceable to a 1952 *Scientific American* article.

With regard to the first meaning, in technical writing today the term has easily grown beyond manufacturing, which was the context of the first use. The second meaning is probably the most widely accepted today, but in this case there is a restriction to what is commonly meant by *automatic control,* which is discussed later. The third meaning, however, is now gaining wide acceptance — namely, any mechanical or electronic replacement of human labor, where labor is taken to mean either physical labor or mental labor. Physical labor was of concern decades ago; mental labor is of primary importance today, at least in the developed nations. Computers that interpret inputs, record data, make decisions, or generate displays are now regarded as automation, including the sensors that go with them, even though in the strict sense none of these functions may be automatic control.

In the fullest contemporary sense of the term, *automation* refers to (a) the mechanization and integration of the sensing of environmental variables (by artificial sensors); (b) data processing and decision making (by computers); and (c) mechanical action (by motors or devices that apply forces on the environment) or "information action" by communication of processed information to people (see Figure 1.3). It can refer to open-loop operation on the environment or closed-loop control.

The term *computerization* formally refers to the instrumentality (the computer) by which a data-processing function is performed. Insofar as the data-processing function is one that replaces (or could replace) a human, one can say that computerization is automation. Computers can perform

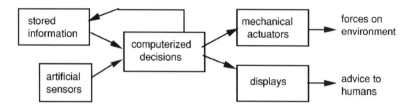

Figure 1.3 The scope of automation.

classical feedback control, which will be considered next, or they can implement crisp or fuzzy rule-based control, which are discussed in Appendix A.

History of Automation

Historically, the first automation was mechanization of manual labor during the Industrial Revolution, although the term was not in use at that time. Steam engines and, later, electric motors were attached to mechanisms that performed routine operations (e.g., sewing, metal stamping) open loop (i.e., without feedback control or self-correction of errors).

Automatic control (feedback control) devices existed early, the most celebrated example being the flyball governor (left panel of Figure 1.4), which was developed to regulate the flow of steam into a steam engine. As the engine shaft (1) rotates faster and faster, centrifugal force acting on the flyballs pushes the balls out (2), moving against a spring, and making the collar slide axially along the shaft (3). This motion is attached to the steam entry valve in such a way that as the balls spin faster, the valve closes more, thereby reducing steam flow to the engine and tending to reduce the

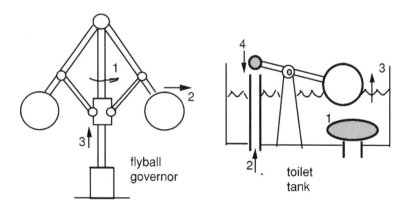

Figure 1.4 Two early automatic controllers.

shaft speed. In this way, with a given setting of the flyball spring, the steam engine comes to an equilibrium speed. Hooking up the flyball device so that faster shaft motion (and flyballs pushing farther out) closes the steam valve is called *negative feedback*. If the attachment were made in the reverse manner, such that faster speed would open the valve, the steam engine would go faster and faster until either it destroyed itself or its speed became limited by its own internal friction.

Another classical example of automatic negative feedback control is the toilet tank (right panel of Figure 1.4). After the toilet is flushed by opening the main valve seat (1) and the water flows into the bowl, the main valve seat (1) then closes by gravity. Then the water supply pressure forces water into the tank (2) and the level rises. As it does so the float ball (3) goes up and eventually forces the supply closure valve (4) down, thereby shutting off further flow into the tank. In this way the water level is regulated to a particular depth in the tank.

Driving a car is another example of negative feedback control, although in this case it is manual rather than automatic. The driver senses whether the car is to the left or to the right of where he or she intends the car to be on the road. If it is to the left, the driver turns a little to the right, and vice versa (this is the negative feedback). In other words, the driver's output — to null any error between desired and actual position on the road — is negative to the visual input. The control gain is the driver's sensitivity to error, the size of the negative steering response relative to the size of the sensed lateral error. The larger the gain, the less the car will deviate owing to any disturbances. (However, if there is a time delay in the system and the gain is too large, control can become oscillatory and even unstable; see Appendix A, section 5.4, to understand why.)

The history of automation includes many engineering developments of sensors, actuators, computers, robots, and all the associated physical systems.

The Status of Automation Today

Automation technology has grown sophisticated, driven by market demands as well as theoretical developments that have occurred at a great pace. Although the diversity of automatic control techniques in the literature is extensive, including so-called optimal control, adaptive control, fuzzy control, and neural nets (the first three are reviewed in Appendix A), the vast majority of practical control applications employ the 50-year-old classical control techniques (see Appendix A). A good general reference to control instrumentation is Shell and Hall (2000).

Computer chips are now part of automobiles, medical devices, home appliances, cameras, watches . . . the list is endless. All of these could be called "automation" in terms of the third Oxford meaning cited earlier. Automation is now a major part of engineering analysis and design, crossing into every traditional branch of engineering.

A Preview of the Critical Issues of Humans and Automation

In this section some hints are given regarding what humans and automation is all about. The point has already been made that automation is here and will remain. In fact, it has been a kind of technological imperative — design engineers automate because they can — and is not always a good idea.

Humans, too, are here to stay — at least as long as the automation, I hope. The automation is supposed to serve the people, not the other way around. So the two must get along. They must live together. They must work together. They must interact.

Automation is still foreign to most people, though. They don't understand it. The more sophisticated it gets, the less they understand it. When they don't understand it, they may not trust it. Or they may overtrust it, attributing to it intelligence that it really does not have. Automation is silent and opaque. It does not reveal its intentions. The people around it cannot always predict what it is doing at the moment or what it is going to do next.

Automation is mostly stupid and single-minded. Unlike people, it is not robust and adaptable. It does what it is programmed to do, which is not always what is desirable or even what the humans using it or affected by it expect it to do.

However, automation can be made to operate in different ways, provide different displays, be controlled in different modes. The problem is that people forget which display or control mode they set it in. Then they may misinterpret what they think the automation is telling them, or what the automation does in response to their command may differ significantly from what they think they asked it to do.

Automation can be reliable, and when it is, people become bored trying to monitor it. People are really not good monitors, anyway. Sometimes the automation fails, though, and when it does, it may be very difficult for the operator to wake up, figure out what has failed, and take corrective action. There simply may be too little time for the human operator to adjust before serious consequences occur.

One form of automation is the decision aid, the advice giver, the expert system, the management information system. It doesn't act, it just tells people how to act. The myth is that decision aids are always safer than automatic controls because the people can still be depended on. The problem with such systems is that people come to trust them and then to overtrust them. Then the people tend to cease acting independently. At that point the decision aid might as well be connected directly to the machine — the human has no apparent function anymore.

Introducing automation can also be intimidating and alienating. A worker in a manufacturing plant who previously had been a skilled craftsman becomes a button pusher — at least, that's the way he sees himself. He may come to feel that the automation is really in charge, that he has no particular contribution to make, so why should he accept responsibility? The designer of the automation is the only human who is responsible.

It should be evident from these thumbnail expressions that automation poses problems, real and imagined, for the human user or other persons who may be affected. The remainder of the book deals with these issues.

Chapter 2

HUMAN INTERACTION WITH AUTOMATION IN VARIOUS CONTEXTS

This chapter examines various contexts in which automation, accompanied by human interaction, has already found wide application and acceptance. This is not to say that automation in these applications is an unabashed success — there are plenty of problems. The applications discussed here are probably the areas where automation is occurring at the greatest rate and where there is the most debate about what should be automated and what should be left to the human operator. The problems discussed under each application are not a comprehensive list, only a sampling. Perhaps readers will begin to see that generalizations can be made that are common to all applications. These will be reviewed in Part II.

Aircraft and Air Traffic Control

Compared with other human-machine systems, it can be said that aviation exhibits the most extensive degree of automation, mostly in the aircraft themselves. Automation within the aircraft takes many forms. Automation within commercial aircraft has received the most attention because it is there that public safety is at greatest risk. Automation for private aircraft has tended to follow behind that for commercial aircraft because of economic constraints in the consumer market. Military aircraft differ in essential respects: they are designed for performance, not efficiency, and they carry weapons. More recently, air traffic control has been undergoing extensive automation (Wickens, Mavor, Parasuraman, & McGee, 1998).

In spite of these advances in automation, aircraft accidents continue to be attributed mostly to humans. The National Transportation Safety Board (NTSB) attributed 144 hull losses (a circumlocution for serious airplane crashes) in 1988 through 1997 to the following categories (Stauch, 2000):

crew error	105
aircraft failure	15
maintenance failure	9
weather	7
air traffic control	5
other	3

Aircraft Piloting

The so-called glass cockpit was introduced in the 1970s with Boeing's 757 and 767 commercial airliners, in which information that previously was presented on separate displays was integrated by computers and displayed on several cathode ray tubes (CRTs) or light-emitting diode displays (LEDs). These replaced the multiple independent mechanical flight instruments and have permitted simplification of the instrument panel. A plan-view electronic map display that shows flight plan route, weather, and other navigational aids is illustrated in Figure 2.1.

Other recent technologies for commercial aircraft (Billings, 1997; Sarter & Amalberti, 2000) include the Traffic Alert and Collision Avoidance System (TCAS) based on forward-looking radar, the Global Positioning System (GPS), and various aids to detect wind shear (in which neighboring air masses have significantly different velocities and an aircraft can suddenly lose altitude).

A recent addition to the flight deck is the Flight Management System (FMS). The FMS extends a trend in aircraft control that began with purely

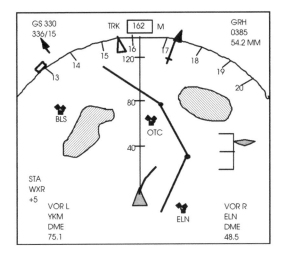

Figure 2.1 Generic plan view display integrating heading, track speed, way points, radio frequencies, course prediction, weather, and other information (after Billings, 1997).

manual operation. Originally manual control was through a two-axis stick to control ailerons and elevators. The stick later became a yoke, whereby the left-right axis of a long stick control between the pilot's legs became a steering wheel. Pedals controlled the rudder and brakes. The manual control became power-aided (like power steering in a car) and still later became fly-by-wire (with no mechanical connection to transmit forces and displacements). Purely manual control was replaced or extended in capability, first by simple autopilots to hold an altitude and/or heading, later by improved autopilot modes to achieve an altitude or heading in a transient maneuver, and finally by inertial guidance modes to automatically take off or land the aircraft or go to a designated navigation way point. One can think of this as a hierarchy of control modes, each with a successively larger scope and time characteristic, all of which are still available to the pilot within the FMS.

At first there were few buttons to push, but gradually piloting has become, to a far greater extent, a matter of button pushing. This trend started as a matter of pushing buttons on the panel that were uniquely dedicated to specific functions, many of which are still there, now mostly collected into a mode control panel. However, the pilot increasingly uses the general alphanumeric key set of the control-display unit (CDU). Typically one CDU is at the most convenient location for the captain's right hand, and another is similarly located for the first officer's left hand, in such a way that either pilot can back up the other. There is still need for some manual control, and in some aircraft only a small side stick remains for either pilot. In the Airbus A-320 the primary flight mode is fly by wire through the small side sticks, in dramatic contrast to the earlier control yokes. The space directly in front of the pilot has become a desk, and now the pilot is sometimes called a "flight manager"!

The current Flight Management System is the aircraft embodiment of what I later (Chapter 6) refer to as the *human-interactive computer:* a computer system designed to talk to the pilots in their own language and to interact with multiple special-purpose control computers scattered throughout the aircraft. The FMS permits pilot selection among many levels of automation, provides the pilot advice on navigation and other subjects, and detects and diagnoses abnormalities. The typical FMS CDU has a CRT display and both generic and dedicated key sets. More than 1000 modules provide maps for terrain, navigational aids, and procedures. A closely associated Engine Indicating and Crew Alerting System (EICAS) presents synoptic diagrams of various electrical and hydraulic subsystems. This computer-based expert system gives the pilot advice on engine conditions, how to save fuel, and other topics. Performance management systems are now available to optimize fuel and time. All of this added display-control automation on commercial aircraft has enabled transition from the three-pilot crew of 25 years ago to the two-pilot crew of today. In response to a given flight plan, the FMS can automatically visualize the trajectory and call attention to any way points that appear to be erroneous on the basis of a set of reasonable assumptions. (This might have prevented the programmed trajectory that allegedly took KAL 007 into Soviet airspace in 1983, where it was shot down as a suspected hostile intruder.) However, it has also led to overtrust and confusion, and in one dramatic instance helped cause a fatal accident (over Cali, Columbia, described in more detail later). Table 2.1 lists some of the functions of the Flight Management System.

The move from control sticks and yokes to button pushing is a move away from direct body image control (sensory-motor skill) to supervisory control through a computer intermediary (cognitive or management skill). It is the FMS, more than any other technology, that forces this move because it dictates that the language of communication (and probably more critically, the language of the pilot's thinking) be alphanumeric (words, mnemonic codes). The pilot's mental models still include the physics of the aircraft — that is, the lift, drag, thrust, and wind components of force that propel the aircraft in its Newtonian trajectory — but must now also include more of the "if-then" logic that computer technicians use.

In spite of the automation being provided, the pilot is still a very busy human operator. For example, during approach to landing in a rainstorm, the pilot may have to perform the following tasks (Adams, Tenney, & Pew, 1995):

TABLE 2.1
Functions of the Flight Management System

Route planning and navigation advice
    weather patterns
    heading to way points for given winds
    speed and fuel burn at different altitudes
    maps and GPS localization
Airport traffic patterns, runway and taxi configurations
Status of fuel, engine, other internal systems
En route traffic collision avoidance
Alarm diagnostics
Digital data link to ground
Flexibility of display modes
Multiple modes of autopilot command

- monitor the descent (compare actual with desired flight path, airspeed, etc.);
- perform the prelanding checklist;
- set the flaps/slats;
- receive radio messages from air traffic control to reduce speed and watch for traffic;
- enter the new speed restriction into the mode control panel (or equivalent);
- look out the window for traffic;
- respond to air traffic control about traffic, seen or not (*seen* means the pilot assumes responsibility for separation);
- worry about the presence of virga, lightning, and dust rings;
- watch airspeed , especially for evidence of wind shear encounter;
- ignore radio communication to other aircraft, except when it contains warnings of nearby hazards; and
- monitor the copilot's performance.

Obviously, even given sophisticated automation such as the Flight Management System, the pilot is still busy and prone to make errors. For example, in December 1995 an American B-757 crew approaching the Cali, Columbia, airport became confused over navigation fix abbreviations that were very similar in appearance to one another and allowed the aircraft to wander off course — unfortunately, across a mountain from the long, narrow Andean valley where the crew thought they were. They finally set the FMS in a "direct to" mode, which deleted the intermediate navigation that might have alerted them to their current location but instead caused the aircraft to execute a turn and crash into a mountain.

An instance of automation overtrust occurred in 1994 as a Jetstream 41 business jet was making an automated instrument landing at night. The first officer was new, and the captain, who was at the controls and said to have had a weak record, was making a perfect descent, except that his airspeed was too low. Allegedly the captain assumed that the airspeed was under the control of the autopilot, which was not the case. The aircraft lost altitude and crashed.

Naturally, the diversity and sophistication of new automatic systems on the flight deck have been accompanied by many new system states and control modes for the pilot to keep track of. There is serious concern about pilot confusion over modes and states in automated cockpits (Degani & Wiener, 1997; Dekker & Hollnagel, 1999; Sarter & Amalberti, 2000; Sarter & Woods, 1997; Sarter, Woods, & Billings, 1997).

Air Traffic Control

Control of the aircraft is determined not only by the pilot but also by air traffic control (ATC). The same kinds of technological changes that have been taking place in the aircraft itself have been happening in ATC. Traditionally ATC is accomplished through a network of stations around the world that are connected to the aircraft by two-way radio and that "see" the aircraft by means of radar. In most countries, both commercial carriers and general aviation aircraft are required to carry transponders that identify them to ATC with a simple code. If the pilots plan to fly into the airspace of major airports, these transponders must have the capability to transmit aircraft altitude as well. This provides an identification tag next to the blip seen on the ATC operator's radar display. Figure 2.2 shows a typical ATC radar display.

Airports with control towers (the smaller general aviation airports are an exception) typically have one ATC operator for takeoffs and landings and one for ground (taxiway) operations, both of whom operate using visual observation. Somewhat larger airports have an additional operator for approach and departure operations, who works primarily from a radar screen. The most critical operations are usually found in the terminal area

Figure 2.2 ATC radar display. (Photo courtesy of Raytheon Company, Lexington, MA.

radar control (TRACON) rooms associated with major airports. Here, the ATC operators must ensure that aircraft arriving from various directions are ordered with proper spacing into the landing pattern and out of the way of other aircraft that are leaving that airspace or that are just passing through. A final category of ATC operators is found in the en route facilities (one for each major sector in a large country such as the United States). These monitor commercial traffic on major airways (so-called highways in the sky) laid out between radio navigation beacons. Some recent technology permits every aircraft in the United States to be displayed on a single large screen in a national flow control center. During major weather problems — for example, when airports must be closed — it is important to monitor and control the cascading effects of backups and delays from one sector to another.

The Center TRACON Automation System (CTAS) is typical of new automation to help the air traffic controller maintain required separation between aircraft while ordering them into proper sequence and runway and providing speed and descent operations. CTAS consists of three separate tools: a sequence and landing time adviser, a descent speed and altitude adviser, and a final approach spacing adviser. CTAS also contains look-ahead conflict probe capability, taking into account flight plans, winds, and aircraft speed. Although there is no doubt that CTAS can make the controller's instructions to the pilot closer to optimal, there is at the same time fear that the human controller will come to depend on the computer-based tool and lose his or her own skill in the process.

Another new automation system is the User Request Evaluation Tool (URET), which helps en-route controllers to anticipate separation conflicts. It is essentially a conflict probe that graphically displays the predicted positions of any aircraft 20 minutes into the future based on the most likely scenario (flight plan). It also is a planning tool that enables the controller to evaluate the results of different instructions that might be given to the aircraft. These various new automation tools are reviewed in Wickens, Mavor, et al.(1998).

The reader has no doubt noticed already the plethora of abbreviations and acronyms, which is typical of the aviation community (see Wickens, Mavor, et al., 1998, for a five-page list of aviation acronyms). It is interesting to note, in the case of the abbreviation CTAS, that the *T* stands for TRACON, itself an acronym; it has largely been forgotten that *radar* is also an acronym, referring to *ra*dio *d*etecting *a*nd *r*anging. Thus CTAS is a third-order abbreviation!

New and better weather radar, weather prediction technology, and communication links between ATCs and supporting weather and other facilities have evolved. As international air travel has increased, the prob-

lems with communication have required rulings that pilots and ATC operators speak a common language, which usually is English. In practice,
however, these rules are not always adhered to. In remote areas controllers
and pilots who share a non-English native language often (illegally) defer
to their own language over the radio.

Heretofore communication between aircraft and ground has been by
two-way voice radio only (except for the aircraft identification transponders). Radio is limited because only one person can talk at a time,
messages often get cut off, and people tend to hear what they expect to
hear. The most serious accident that ever occurred — two 747s crashing on
Tenerife in the Canary Islands — was attributed to incomplete and misunderstood radio messages combined with visibility problems. One aircraft
was just at takeoff, the pilot thinking he had proper clearance, while the
other aircraft was taxiing up the runway after landing and did not turn onto
the taxiway. There was intermittent cloud blowing across the runway,
hampering visibility.

One major innovation is data-link, a two-way communication between
aircraft via digitally encoded messages. Data link allows much more data
communication with the ground than is possible with traditional voice
channels. It enables new capabilities but also poses new problems.
Although its capabilities include sending a variety of information about
aircraft conditions to the ground for recording and analysis in real time,
commercial rights to privacy inhibit full use of this particular function.
Data link capabilities also include making available to the pilot any or all
of the radar information about other aircraft in the pilot's area, beyond the
short-range radar that TCAS provides, which might seem to be an advantage. The problem is that if the pilot sees all of what the ATC operator sees,
and the pilot is ultimately in charge of the aircraft, does ATC thereby lose
control?

Perhaps even more serious is the question of how communication
between ATC and the pilot is accomplished. With data link, ATC instructions, landing clearances, and so forth can be sent to the aircraft digitally
and be put on visual displays, and pilot queries and responses can be sent
to the ground digitally, thereby (perhaps) obviating the need for voice
communication. However, pilots like voice communication and have
confidence in it. It is immediate, it is flexible, it can be as informal as
necessary, and it offers immediate confirmation that the other party is
paying attention. Further, by listening to the "party line," pilots can gain a
sense of what is going on in the surrounding air traffic. Communicating by
data link may be more reliable in a narrow technical sense (voice messages
get cut off and essential information can be lost, require repeating, etc.),
but data link communication could turn out to be less reliable in a broader

sense. What about having both? That may increase workload. So the form in which data link will ultimately be implemented remains under active discussion and research.

Free Flight

Because of much-improved location and separation capability through GPS, TCAS, and related sensor and communication technology, the pilot presumably is potentially less reliant on air traffic control now than in the past. This has given rise in the United States to the idea of *free flight*, under which aircraft would be free to fly en route by direct great circle routes and to deviate from those as they wish, either vertically or horizontally, depending on winds aloft, weather conditions, and other air traffic, all of which they now can observe for themselves. Horizontal free flight would reduce time and fuel costs considerably, as compared with having to chart all flight plans between way points, which are not in a straight line. The pilots still would have to abide by ATC instructions as they converged on their destination airport. Vertical free flight would allow the pilot to access whatever altitude nets the greatest advantage with respect to winds and weather. Initial studies suggest that significant fuel savings are possible by using decision aids for vertical free flight (Patrick & Sheridan, 1998).

Military Aviation

Military aviation has many of the same advances and problems of automation as does civilian aviation, but it has the added factor of dealing with an enemy. In fighter aircraft, including helicopters, there are many highly automated weapons systems, including the associated radar and optical systems for both weapons firing and protection of the aircraft by maneuvering.

Unmanned (unoccupied) air vehicles, or UAVs, are coming into wide use by the military services for reconnaissance and weapon delivery. The UAV is remotely controlled by a pilot on the ground. Such piloting is very different from piloting within the aircraft because vestibular and other motion cues are lacking, vision is via television, control is more supervisory than direct, and all the human-machine problems of teleoperation (to be discussed later in more detail) emerge.

Passenger Interactions

Automation affects not only the pilots and air traffic controllers but also the passengers — for instance, in ticketing operations, baggage handling, security screening of both passengers and their carry-on luggage, and getting from terminal to gate and the reverse. Passenger delays in all these operations can add up, sometimes exceeding the time for the flight itself. It seems there is still room for improvement by using or refining automation, especially to reduce delays in baggage handling.

Automobiles and Highway Systems

Highway vehicles — including private automobiles, trucks, buses, taxicabs, and special vehicles such as fire trucks, police cars, and ambulances — have also followed the trend toward use of new sensor, communication, computer, and control technology. The goal is "intelligent" transportation systems (ITSs) to improve both safety and mobility. The European Community countries were probably the first to organize a major development project. Two such projects, called PROMETHEUS and DRIVE, have been completed. Their aims were to introduce new computer-based navigation and control systems; to look ahead to common signing, speed control, and other highway infrastructure developments for European countries; and to enhance the position of European automotive manufacturers in world markets. In Japan similar national projects instituted cooperation among vehicle manufacturers, suppliers, and government agencies.

In about 1990 the U.S. Department of Transportation, in concert with state agencies, vehicle manufacturers and suppliers, insurance companies, and universities, founded a unique advisory and educational organization called the Intelligent Transportation Society of America (ITSA) and began funding research and development on "smart" cars and trucks. From the outset, human factors considerations were seen as important to the success of ITS. A review of current human factors problems with highway transportation is found in Barfield and Dingus (1998).

Navigation
Also in the early 1990s General Motors, Avis, the American Automobile Association, the State of Florida, and the City of Orlando undertook a demonstration project in passenger vehicle navigation. In this project 100 rentable cars were outfitted with GPS transducers, simple inertial transducers, PCs, CD-ROM readers to hold street map information for Orlando, speech generation devices, and CRT graphical display units. The rental agency customer could indicate his or her destination to the computer, and the system would then guide the driver by a combination of voice and graphical displays. This included, for example, advice to move to one side of the road or another in preparation for a left or right turn at a forthcoming named intersection. The GPS was accurate enough in most cases to detect whether the driver stayed on the correct road, and if not, it would tell the driver how to get back on the correct course. From this first human-

automation demonstration a variety of navigation systems evolved that are now available commercially. Controversy over putting such navigation technology into passenger cars mostly centers on safety issues. Does use of such a system result in greater confidence and safety, or does it detract significantly from the driver's attention to the road? Can older drivers accommodate their visual focus sufficiently rapidly between the road outside and the internal graphics or other displays? Is this yet another addition to the driver's workload? Are map displays too distracting? (Some U.S. states claim they are and legally prohibit front-seat map displays, classifying them the same as television sets.) If maps are used, should the map direction always correspond to the vehicle direction (the vehicle then becoming a fixed symbol), or should the map remain fixed with north (or south) up, in which case the vehicle becomes a moving symbol? The answers to these questions may be settled to some extent by research, but the final arbiter will be the marketplace. See Eby and Kostyniuk (1999) for a comparison of various car navigation systems in on-road tests.

Collision Avoidance

The major safety objective is to avoid collisions. New technology provides a different approach to collision avoidance. Instead of accepting the first crash of the vehicle with environmental obstacles and focusing only on technological means to ameliorate the second crash of the driver and passengers with the interior of the vehicle, technology now has the opportunity to prevent the first crash. In this regard, perhaps the second major new development for intelligent vehicles is *intelligent cruise control.* Current cruise control systems do not "know" when one vehicle is about to collide with another. Intelligent cruise control systems make use of microwave and optical sensors to detect the presence of a vehicle in front, and the resulting signal can be used either to warn the driver or to brake automatically.

Sadly, there are some troublesome problems with this approach. These are illustrated in Figure 2.3. The radar beam width must necessarily be restricted so that it does not pick up vehicles or objects other than those being followed in the same lane. Unfortunately, when a following car A begins to overtake a lead car B (see upper part of Figure 2.3), it must turn so that the radar beam no longer picks up the lead car. Then the following car can suddenly speed up unnecessarily unless there is intelligence to prevent that. Further, on a curve the beam from a following car C may not pick up the lead car D at all (see lower part of Figure 2.3) unless there is intelligence to actively track that car. But such intelligence must also go back to straight ahead if the lead car turns off the given road.

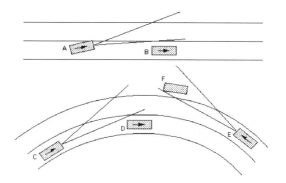

Figure 2.3 Problems encountered by early prototypes of intelligent cruise control.

Finally, a car E on a curve may inadvertently pick up a car F parked off the road unless there is intelligence to recognize that car F is off the road. A suggested solution for this problem was to ignore all objects that are closing at the same velocity as the parked car, which assumes that it is off the road. But what if it is actually on the road but just stalled? Clearly, intelligent cruise control must be more intelligent than the initial prototypes, which could not cope with these problems.

Interestingly, initial demonstration projects on intelligent cruise control mostly stayed away from actually applying the brakes, choosing instead to decelerate and downshift. The reason for this seemingly irrational choice is the concern that once claims are made that an automatic system will brake, the driver may become less inclined to apply the brakes manually, leaving the manufacturer wide open to tort litigation.

In simulator experiments to test braking reactions and following distances of drivers under various conditions, Chen, Sheridan, Kusunoki, and Komoda (1995) found that following distance correlates surprisingly poorly with the speed of the traffic stream, to the point that most drivers could not stop in time to avoid collision if the driver of the vehicle ahead were to apply the brakes suddenly. From varieties of such empirical data, one can run a Monte Carlo quantitative model by assuming the initial speed of the traffic stream, the stopping distance of the lead vehicle, the braking characteristics of the following vehicle, and the pavement conditions, either random or biased. Draws can be made from density functions (obtained experimentally) for the following-car driver's reaction time and following distance. This can then be used to calculate the probability of collision.

In 1994 the U.S. Department of Transportation initiated a major program to test automated highway technology. European and Japanese companies were invited to participate, and they accepted. In San Diego in the summer of 1997, many new automated vehicle control techniques were demonstrated, such as automatic lane following based on magnets buried in the highway or light-reflective strips; automated platooning of cars with automated electronic separations of only a few meters, based on radar. In one sense the demonstrations were a great success. However, it was clear that many surrounding problems had hardly been touched on — for example, how the computer takes control from the human as the car enters an automated highway, and how it gives control back if something fails. There was also the problem of how vehicles without automation should or could interact with those that have it. The problems of authority and responsibility among the driver, vehicle and vehicle equipment manufacturer, and local governments (which might be responsible for highway infrastructure maintenance), as well as who might litigate against whom for what, loomed large, especially in a litigious society such as the United States. It was clear that technology problems were easy compared with the human problems. Consequently the project was discontinued.

There is much concern about driver distraction caused by increased use of cellular telephones while driving and about the potential for the driver using e-mail, faxing, watching television, and using interactive entertainment devices. No vendor of such technology wants to deal with negative concerns, and neither vendors nor government agencies want to be responsible for the interactive effects that new devices have on each other in terms of competing for driver attention.

Advanced Traffic Management

The intelligent highway system also has its counterpart in air traffic control, called the *advanced traffic management system.* This is not a unique concept; in fact, many new highway and tunnel projects have built accompanying traffic management centers. The control is limited; the driver is still free to a large extent. Murray and Liu (1997a, 1997b) characterized central control as *hortatory* and offered a means to differentiate demand-based congestion from incident-based congestion.

The newest such system is associated with the largest current construction project in the United States, Boston's $12 billion Central Artery/ Tunnel. It includes 60 lane kilometers in multiple tunnels. In this project's control room, several operators observe traffic by means of roughly 400 video cameras and a comparable number of magnetic and other sensors of traffic, heat, smoke, carbon monoxide, and toxic chemicals of various kinds. In case traffic becomes halted in a major tunnel, toxic fumes can exceed safe levels within an hour, despite large blower installations. The new Boston tunnel uses a computer system that infers where and how serious any abnormality is and makes a recommendation for action (e.g., send a tow truck from a particular location, send certain firefighting equipment or police). The operator can change variable message signs, block traffic lanes, initiate visual and auditory alarms, and activate fire extinguishers.

With regard to such emergency response situations, one supervisory control problem that has not received the research attention it deserves is *intervention*. In the Boston system, the supervisory operator is expected to decide within seconds whether to accept the computer's advice (in which case the communication to the proper responding agencies is automatically commanded) or to reject the advice and generate his or her own commands (in effect, to intervene in an otherwise automatic chain of events). Human-in-the-loop simulation of such emergencies is needed to evaluate what one can expect of a trained operator under various crisis conditions.

Trains

Although the aircraft is surely the most high-tech mode of transportation − as well as the newest, the one with tightest selection and training requirements for human operators, and the most regulated in other respects − it is interesting to note how automation changes are occurring in older transport modes. Consider railroad trains, a transportation mode much older than aviation or highway vehicles. Though slow to follow, railway systems are copying much from aviation technology, such as the use of the GPS for determination of latitude and longitude and the use of new sensing technology to measure critical variables (e.g., the temperature of wheel bearings). Locomotives are even using fly-by-wire controls, which, in the case of trains, combine the brake and throttle adjustments into a single control. (Existing systems have separate controls for the brake and throttle, which evolved from air pipe valves and locomotive steam valves that the driver used to operate directly.)

The Train Driver's Task

A train driver's primary task is to control speed and, as part of this, to anticipate the uphills and downhills that tend to decelerate or accelerate the train by gravity. The driver must know the fixed speed limits at different locations along the track where there are curves, switches, grade crossings, or heavily populated areas. He or she must stay current with information about temporary speed limits arising from weather conditions, problems with the track or roadbed, or maintenance crews working at certain locations. The driver must know the train momentum and friction effects of the wind and track as a function of speed for different consists (number of cars and type of load). He or she must know the limits of train thrusting and braking (both normal and emergency).

The driver is responsible for reading and comprehending the wayside signals at the entry to each block or section of track, which becomes more difficult as speeds increase. He or she must communicate with the central dispatcher as necessary regarding unexpected train movements. The driver must know, or be able to access, the scheduled time of arrival at each station and be able to brake the train to a stop at precisely the correct point at each station. The driver must also monitor brake pressure and other variables of train and track condition, change the pantograph connection to the overhead electrical power cable from DC to AC and reverse, and operate diesel motors, as required. The driver must integrate all of these so as to maintain safety, arrive at the next station on time, and minimize the use of fossil fuel or electrical energy.

In some sense the train driver's job looks easier than that of the airplane pilot, in that one might think of a train as a one-dimensional airplane (the train is confined to a track; the aircraft flies in three-dimensional space). The special problem with a train, however, is that it has very large momentum (it takes up to 3 km to stop a modern high-speed train traveling at 300 km/h, even under emergency braking). If an obstacle lies ahead — for instance, a truck stalled on a grade crossing, a bridge or section of track that is not in good condition, or a large rock or rockslide from a mountain — the train does not have the maneuvering room an aircraft has; it must stop.

Some trains are equipped with automatic speed controllers that can be set, much like cruise control systems in automobiles. However, similar to the automobile's cruise control, this does not take care of necessary braking. Some newer trains do have automatic train protection or positive train control systems that signal the driver to put on the brakes if another train is detected immediately ahead (trains have been easier to detect than other obstacles by means of electromagnetic sensors) and that apply the brakes automatically if the operator does not respond in time. However, the existing speed signals, which are mostly on fixed posts along the side of the track bed, are not only imperceptible from a distance but are also increasingly difficult to read as the trains reach faster speeds.

New Displays

Flexible, computer-driven displays for monitoring that are located on the aircraft pilot's console (the glass cockpit) are gradually finding their way into railway systems. What might such displays be used for, and how might they assist the driver in speed control? One obvious use is for displaying, in the cab, the same information that now must be gleaned from the outside signals and providing this in an easy-to-read manner right in front of the driver. This would depend on digital communication between the wayside and the locomotive cab, which would mean an expensive overhaul of the present signaling system. Even though speed signal information from the wayside is being transferred into the cab itself, this does not help if the operator cannot see trouble far enough ahead. The driver certainly cannot see more than 1 km ahead through the windscreen and can see much less than that at night. (Headlights on trains are a signal to outsiders and provide little illumination for the driver.) Experienced drivers use a mental model of speed constraints set by curves in the track, grade crossings, and population densities, which are fixed and can be learned. As noted earlier, though, because of track maintenance activities, rockslides, snow, and so forth, there may be other speed constraints that are not as easily anticipated.

An advanced version of an in-cab display has been proposed (Askey, 1995), as shown in Figure 2.4. Its purpose is to help the driver anticipate by previewing speed constraints, predicting the effects of alternative throttle and brake settings that might be set momentarily, and providing an optimal throttle setting. It also shows prediction curves of the speed, determined from a computer-based dynamic model. These curves show how the

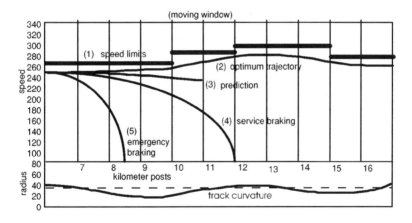

Figure 2.4 Askey's experimental preview prediction/adviser display. At the right, referenced to the speed indicator (vertical) and distance from the present position (horizontal), are (1) bars at the top ("preview"), indicating speed limits for upcoming kilometer sections; (2) upper curve ("adviser"), indicating optimal speed trajectory to satisfy speed limits, schedule constraints, and energy minimization; (3) curve almost coinciding with first part of the upper curve ("predictor," displayed in contrasting color), which indicates prediction of train speed if current throttle setting is held constant; (4) next curve ("predictor"), indicating maximum service braking; and (5) lower curve ("predictor"), indicating maximum emergency braking. The curves are continuously recalculated based on present state and computer models of train dynamics and track configuration.

speed will change as a function of the track distance ahead if the current throttle setting is maintained. There are also predicted speed curves for maximum service braking and maximum emergency braking (different braking systems). Finally, there is a continuous indication of throttle settings that will get the train to the next station on schedule (assuming the train is at or near schedule currently and that the winds are known), meeting the known speed limits, and, under these constraints, minimizing fuel consumption. This indication can be updated iteratively by a dynamic programming algorithm. The latter display is akin to the flight director in an aircraft. This system was tested in a dynamic human-in-the-loop train simulator with a number of trained drivers and was shown to improve performance significantly over driving with conventional displays (Askey, 1995).

Operator Takeover Policies
A fatal accident on the Washington, D.C., Metrorail (public transit system) several years ago resulted from an unclear management policy regarding driver takeover in case of an automation failure. The system had been set in automatic control by the operations control center, even though some operators had requested a manual control mode because of icy track. However, one new operator, when he found his train was not slowing as expected, was intimidated by what he perceived to be orders not to countermand the automated braking system. His train overran the Shady Grove Station by 470 feet (about 143 m) and struck another train at full speed.

Train Dispatching
Rail systems have their equivalent of air traffic control in the dispatching or traffic control centers, usually located at major rail junctions. Here, a crew of several people set schedules, set switches, communicate with trains by telephone as necessary, and monitor rail traffic for large areas. Until recently their principal tools have been the telephone, large sheets of paper with schedule tables and/or plots of location versus time for every train for various sections of track, and diagrams or map displays, in some cases computerized to show which blocks are occupied by trains. Controllers have had few, if any, tools to predict the cascading effects of late trains or accidents. This situation is gradually changing, and data link technology (as described earlier in conjunction with aviation) is being developed for communications between dispatchers and both train drivers and track maintenance workers (the latter being under some danger if trains do not keep to schedules).

Ships

Surface and undersea vessels are undergoing extensive automation. Modern cruise and bulk carrier ships as well as naval vessels have automatic roll stabilization. Sonar has provided much more reliable tracking of depth and potential obstacles, and GPS technology has provided much more accurate localizing in latitude and longitude for navigation.

However, humans can overtrust automation in ships, just as they can in aircraft and trains (and probably soon will in automobiles). An example is a 1995 incident involving the *Royal Majesty* cruise ship. The ship's GPS antenna had become detached, and as a result the ship went off course. Control was continued automatically on dead reckoning, without the pilot's realizing it. The radar map display showed a course with local features that, to the pilot, appeared similar enough to the desired course that he thought he was entering the desired harbor. The ship ran aground.

Containerization

Containerization is one major form of automation. Shipping containers, typically $8 \times 8 \times 20$ feet (about $2.4 \times 2.4 \times 6.1$ m) are removed from tractor-trailers and stacked onto special cargo ships at dockside, and the procedure is reversed when the ship reaches its destination. In Rotterdam, Netherlands, for example, one finds the largest and probably most automated container port in the world, where loading and unloading is accomplished by programmed robotic cranes. Driverless vehicles, each containing one container, shuttle between shipside and the tractor-trailer area. Because the tractor-trailers differ in size and shape, the loading and unloading must be done by human-operated cranes. The many shipside cranes and shuttle vehicles are programmed and monitored by only a few personnel in control towers, who must always be prepared to take over.

Crew Size

There have been major efforts in recent years to reduce the crew size on merchant and naval ships. Ship power plants and other machinery no longer need a dozen or more personnel for inspection and maintenance. Watchkeeping is accomplished from the bridge with automated sensors and communication. Whereas multilevel chains of command used to be employed to pass commands from officers down to the lowest levels of personnel who performed manual tasks, now communication is sent through computers to sensors and actuators throughout the ship. In the U.S. Navy many jobs were specialized, so sailors who had specialized in manning weapons or in putting out fires were either unemployed or put to work swabbing decks during peace or nonemergencies, while cooks and various technicians kept the ships running. Nowadays, the crew members increasingly are cross-trained in multiple tasks.

Use of Predictor Displays for Maneuvering

The largest ships of all, supertankers, are particularly difficult to steer through harbor channels and into port because of unfamiliar terrain, obstacles, and tides. This is largely because of their huge mass and, hence, the long time constants involved in controlling these vessels. A means to cope with this maneuvering is the use of predictor displays combined with GPS localization. The predictor display is essentially a displayed trace (on a map) of the output of a computer model of the trajectory of the ship based on current position, velocity, acceleration, current, and wind loads, plus a hypothetical program of thrust and steering commands over the next short interval (the interval of the prediction). By observing the predicted trajectory for alternate thrust and steering programs, one can select the best course to take the ship where desired and avoid obstacles. The working of the predictor will be discussed further in Chapter 5.

Sometimes a surface ship needs to tow a submersible at a great depth; for example, to map the ocean bottom with side-scan sonar and photography. For this purpose very long cables are necessary (for towing force and to carry power), up to 5 or 6 miles (8–9.5 km) long for the deepest parts of the ocean. The time delay between forces applied to the top of the cable and those at the bottom can be 10 minutes. For traversing canyons in the deep ocean, the helmsman has an almost impossible problem: Without some predictive device, he or she is blind. In this case a predictor display is based on a model of the ship, the submersible, and the dynamics of the cable interacting with the surrounding water.

Spacecraft

Ever since Sputnik, both occupied and unoccupied spacecraft have been automated to a great extent. The first crew of *Apollo* astronauts, all of whom were seasoned fighter pilots, had to come to grips early in their training with the fact that they were not going to control their craft as they were accustomed to (i.e., with a control stick and pedals). Rather, they would be keying symbolic codes into a computer, and the computer would be controlling the spacecraft. (I participated in training that first *Apollo* crew in the early 1960s.)

More recently, deep-space probes as well as Earth communication and scientific satellites, such as the Hubble and Chandra space telescopes, have been outstanding successes by any measure. These devices have taught much about human supervision from a distance of computer-intelligent devices that have their own sensors and actuators. To realize that humans have been communicating with spacecraft that are nearly one billion miles away is truly astounding.

Doing science in space can be frustrating for the scientist, however, not only because of long time delays (e.g., seconds to the Moon but most of an hour to Mars) but also because scientists have limited control over their experiment, once it departs Earth, because of operational constraints imposed by NASA or another nation's space agency. Even in low-Earth orbit, where an astronaut is assigned to assist with each experiment, there can be serious delay and miscommunication problems if something begins to go wrong. For example, in the preparation of human participants or animal subjects for biomedical experiments, in tasks such as placing electrodes, the astronaut may not be sensitive to problems that arise, and there is no opportunity for the principal investigator on the ground to intervene. For this reason NASA has begun using computer-based expert advisory systems (the rule base programmed by the ground-bound scientist), which are wired into experiments to assist the astronaut in setting up and in diagnosing any trouble.

Another type of human-automation problem relates to the veritable army of personnel that in the past were required to support a staffed space mission. NASA would like to reduce this number through computer-based monitoring and other forms of automation.

Teleoperators for Hazardous Environments, Motion Rescaling, and Other Nonroutine Tasks

A *teleoperator* is a system with artificial sensors and actuators that a human communicates with and controls from a distance. A *telerobot* is a subclass of teleoperator that has a computer with a stored program capable of acting independently (a robot), which a human supervises (communicates with and intermittently reprograms) from a distance. The complementary subclass of teleoperator is a system that is continuously controlled by a human operator. An example of the latter is a position-controlled *master-slave* servomanipulator in which every motion of the human hand and arm is kinematically duplicated (not necessarily on the same scale) by the motion of a mechanical hand-arm. Other examples are a remotely controlled Martian rover, a toy car, or an aircraft operated through a control stick. Many new teleoperator systems incorporate capability for both direct (continuous) and computer-mediated control. Teleoperator technology is reviewed by Vertut and Coiffet (1986) and Sheridan (1992b).

Usually one thinks of teleoperators or telerobots as being designed for space, undersea, mining, toxic, or other hazardous environments. So-called industrial robots in factories are not telerobots. They are generally specialized to particular tasks such as welding or paint spraying, they operate for long periods without human intervention, and the human, if and when interacting with them (e.g., for setup or maintenance), performs tasks hands on, not from a distance. In some sense, airplanes and other systems that are not normally thought of as robots are really telerobots in the general sense, because they embody artificial sensors, actuators, and computer decision making and are subjected to remote (maybe 15 m rather than thousands of kilometers) supervisory control by a human.

Space telerobots have landed on remote planets, roved their surfaces, taken photos, sampled the soil, performed a variety of scientific tests, and returned data to earth. Undersea telerobots have made spectacular discoveries, such as the sunken ship *Titanic* and the so-called black smokers on the deep ocean bottom that expel hot gases and around which all sorts of new life forms have been discovered — suggesting that life can exist at pressures and temperatures never before dreamed of.

Figure 2.5 shows the undersea telerobot Jason, big brother of Jason Junior, which swam inside the *Titanic* after it was discovered. Jason has several propellers for mobility in any direction, a manipulator arm, and a

Figure 2.5 The Jason remotely operated submersible. (Photo courtesy Dana Yoerger, Woods Hole Oceanographic Institution.)

variety of video, sonar, temperature, and other sensors. It can be operated in direct manual mode or in a supervisory mode using sonar triangulation to provide position feedback. Another type of telerobot is that used for tasks such as delivering mail to offices in large buildings or retrieving parts in large warehouses. Yet another type is used for systematically moving across the outside surfaces of large buildings to wash windows or to perform similar functions to clean factory floors. In these cases the buildings themselves must have a cooperative infrastructure to assist in guiding the telerobot, such as optical or magnetic path markers or bar codes to identify rooms and key locations.

Display Aids and Command Language

For the planning phase of supervisory control, a variety of control modes (Brooks, 1979; Yoerger, 1982) and graphical aids (Ellis, 1991) have been developed. Many of them make use in the computer of a geometric model of both the task and telerobot, allowing the supervisor to test a set of commands before implementing them. Various graphical techniques help the user observe the robot arm-hand relative to various environmental objects from any arbitrary viewpoint, observe in the two-dimensional display the orthogonal distances between objects and thereby avoid collisions (Das, 1989), and even get advice from the computer regarding the safest trajectory (Park, 1991).

Command languages include symbolic and analogic means to tell the telerobot how to move. For example, some telerobots are equipped with wheels, tractor treads, or legs for mobility on land; some have propellers for mobility under water; some have thrusters for moving around in space. High-level commands from the human specify task goals (e.g., to pick up a designated object and place it in a new location). "Put that there" has become a common class of commands in which the action command is further modified by descriptors that designate the object to be handled, how to make contact, how to grasp the object, how fast to move, how to assemble, and what should be the final configuration of object and environment.

To complement their remote (tele-) vision sense, teleoperators have also gained a haptic capability. Forces are measured at the remote hand and imposed back onto the human hand that is guiding the teleoperator control handle. Numerous force-feedback devices are being developed, as are means to touch an object remotely and feel the texture (Burdea, 1996). However, touch-pattern display to the skin is a much more difficult problem than simply having the whole control handle push back with the same force that is being applied to the remote environment. Operators of teleoperators, when provided with a head-mounted display or other means to provide good vision of the remote scene, and also provided with good force feedback to their controlling hand from forces encountered there, report feeling a remarkable sense of *telepresence* (the experience of a distant environment as though being there while not actually being there; Draper, Kaber, & Usher, 1998; Sheridan 1992a, 1992b).

Coping with Communication Time Delay

A special problem encountered in continuous manual teleoperation occurs when the communication channel contains a relatively long time delay, such as the 3- second round-trip delay that occurs when controlling a lunar teleoperator from Earth. In this case any attempt to control with a closed-loop gain exceeding unity (normally imperative at frequencies where one wants faithful tracking and noise rejection) will result in insta-bility. A human controller's natural tendency is to resort to a "move-and-wait" strategy, wherein the operator commands a short open-loop move of the remote teleoperator arm, then waits for feedback, then makes another move, and so on. This procedure makes for extremely slow action because the moves must be very short, especially as the teleoperator hand approaches contact with an object. Ferrell (1965), who was among the first to study such human control through a time delay experimentally, constructed a model that predicted task completion time (Figure 2.6). This was precisely the problem that led to the recognition that supervisory control (Ferrell & Sheridan, 1967; Sheridan, 1992b, 1993) — that is, telerobotic control — is both a viable alternative to continuous teleopera-tion and a useful paradigm for human interaction with automation in general. Supervisory control is covered more fully in Chapter 6.

The discrete communication time delays in long-distance teleoperation may also be ameliorated by predictor displays (see Chapter 5).

Virtual Reality for Training and Entertainment

Just as new teleoperation technology enables seeing, hearing, and touching real objects that are arbitrarily distant in space, enabling a sense of telepresence (as defined earlier), virtual reality technology enables seeing, hearing, and touching objects that do not exist at all — or perhaps one should say they exist only as computer programs. Synonyms for *virtual reality* are *virtual environment* and *artificial reality*. From the human operator's perspective, telepresence and virtual reality can look exactly the same (Figure 2.7). Machine-based virtual reality has existed in some form for 50 years (e.g., primitive flight training simulators) but could never be convincingly implemented until very fast graphics engines came

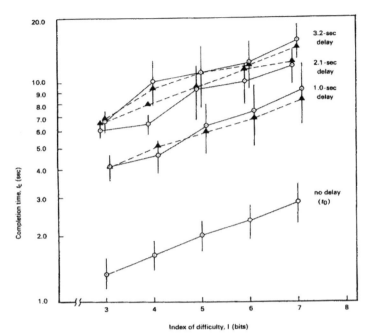

Figure 2.6 Ferrell's results for simple telemanipulation with various time delays. Experiments were performed in two-degree-of-freedom pick-and-place tasks with various accuracy requirements (Fitts index of difficulty) and pure time delays. Open circles are measurements; filled triangles are predictions based on eye closure with each move instead of time delay. Data points are averages across subjects, vertical lines indicate one standard deviation. From Ferrell (1965).

along. (One might argue that non-machine-based virtual reality has been with us for many more years and is provided by storytelling, books, films, television, and the theatre, where the observer enhances the sensation by suspending disbelief that the experience is not real.)

Generating Virtual Environments
The interface technology for virtual reality is critical and is what has made it so popular in such a short time. First there is the visual telepresence technology, typically a head-mounted display and head-position-tracking apparatus that drives the display viewpoint to correspond with the direction the wearer is looking (so that the wearer sees what would be seen if he or she were present in the virtual environment). In teleoperator systems, head motion drives the remote video camera to a corresponding orientation to accomplish the same result. Durlach and Mavor (1995) and Barfield and Furness (1995) provided reviews of this technology.

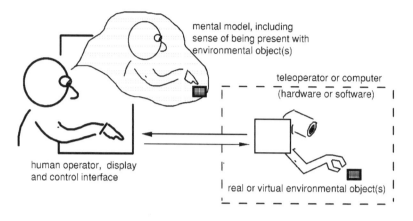

mental model, including
sense of being present with
environmental object(s)

teleoperator or computer
(hardware or software)

human operator, display
and control interface

real or virtual environmental object(s)

Figure 2.7 Telepresence and virtual reality appear the same to the human.

The auditory equivalent of the head-mounted display has also come about in recent years as a result of a better understanding of what the outer ear does: namely, to serve as a nonlinear filter to provide cues of sound source location from front to back and up to down. (Left-to-right discrimination is accomplished by magnitude and time (or phase) differences between sounds reaching the two ears.) By placing a white-noise sound source at different locations and a tiny microphone right on the eardrum, one can determine the *head-related transfer function*. Knowing this, and using earphones on the observer (thereby bypassing the outer ear function), one can simulate various sounds appearing to come from any given direction or having any motion relative to the head.

Haptics is a relatively new area of providing virtual experience. The term *haptics* now refers to the combined senses of forces on the body and limb positions as mediated by receptors in the skin, muscles, and tendons. Haptic experience can be created by having the observer place a finger in a multiple-degree-of-freedom, very high bandwidth motion device that generates a reactive force or vibration as a function of finger position. The experience of touching a surface, scraping one's finger across smooth or rough textures, and so forth is provided and can be very realistic (though necessarily a function of the software that drives the motion).

The body senses motion not only through the muscles and tendons but also through the several vestibular sensors: the otolith organs (linear acceleration) and semicircular canals (angular acceleration) of the inner ear. Because these organs are sensitive mostly to higher frequencies, virtual motions (e.g., the sense of acceleration in an aircraft or automobile) can be generated in a simulator by short-duration motions combined with tilting of a simulator cab. Further discussion of moving-base simulators appears in Chapter 7.

Entertainment is by far the largest market for virtual reality, although skill training (sports, surgery) is catching up. My greatest worry about virtual reality concerns the simulation of violence in computer games and arcade entertainment devices. Worse than violent films or television programs, they enable teenagers to engage in extremely realistic-appearing violence and then immediately walk back into real environments with real people. The tendency could be to seamlessly continue what they have just been doing.

Nuclear Power Plants and Process Control

The term *process control* refers to the control of any physical process that is continuous in time and space, in which the product flows through the manufacturing or transportation operations. This includes electricity generation (nuclear, fossil fuel, hydroelectric, solar, etc.); electricity distribution; chemical manufacturing; petroleum cracking and processing; fabrication of cloth, metal wire and strip, paper, rubber; and so forth.

Process plants are increasingly run by automation. In fact, they were among the first large-scale systems to be controlled this way. Large numbers of simple linear proportional-integral-derivative (PID) controllers regulate flow, pressure, temperature, and other factors scattered about the plant. Set points are centrally adjusted by human operators, who also monitor for many alarms (several thousand in some plants).

Beyond the simple PID controllers, computers have been used in process control for sensory pattern recognition, recording data, performing off-line analyses on data, establishing trends and predictions, and generating displays. In some plants a whole batch run can be programmed for a series of operations as well as various abnormality contingencies.

The Complexity of the Control Room

Over time and with good intentions, the number of independent indicators and alarm lights in the control room has grown into the thousands. In addition, it has become evident that in some plants, after a serious malfunction, too many displays change state. In one nuclear plant large-break, loss-of-coolant simulation, I counted 500 displays that changed in the first minute and 800 more in the next minute. In such events, operators have referred to the control board as a Christmas tree. Under such circumstances there is no way an operator can quickly determine what is going wrong. Experienced operators claim that some obscure form of pattern recognition is the way they try to cope. Figure 2.8 shows an example of a modern nuclear power plant control room.

When the accident at Three Mile Island occurred in 1979, the Nuclear Regulatory Commission mandated a safety parameter display system (SPDS), a synoptic display of the health of the plant — a way to keep the operators informed regarding which major systems were functioning properly and which were not. However, one problem learned by the nuclear power community was that because the SPDS was forced on a community of operators who had little or nothing to do with its invention or dictation of use, it was at first simply ignored by the operators. Technology that purports to be helpful, especially in crisis, must be understood and accepted by the users!

The accident at Three Mile Island ended the American public's confidence in nuclear power, and the much worse accident at Chernobyl in the former Soviet Union seems to have ended the public's confidence within many European countries. Much of the concern is with human error and the human operator's ability to monitor and understand such complex systems, especially when under time pressure. In American nuclear plants there has been reluctance to commit control to computers. They have been used primarily for information acquisition, recordkeeping, warnings, and alarms. In Canada, France, Germany, and Japan their use for control has been much more extensive. Chemical process plants in all developed coun-

Figure 2.8 A nuclear power plant control room: South Texas Project, 1998. (Photo provided by Westinghouse Electronic Company.)

tries have made extensive use of computer control and so-called expert advisory systems.

Gilmore, Gertman, and Blackman (1989) reviewed the special problems of human factors and automation in process control.

Manufacturing of Discrete Products

A *flexible manufacturing system* is a combination of many robotic discrete-parts manufacturing operations (called *cells*) under some combination of automatic (computer-based) and human control within a single factory. A flexible manufacturing system may include one or many computer-controlled cells used to perform individual metal-forming operations (rolling, stamping, extruding, casting, etc.), cutting operations (milling, drilling, boring, reaming), surface-finish operations (e.g., grinding, painting, heat-treating), and the transfer of raw materials for in-process parts between workstations (Figure 2.9). Such a system permits small-batch manufacture and easy change of product without totally restructuring the factory.

Flexible manufacturing system technology is undergoing very active development, and both operations research and optimal-control modelers

Figure 2.9 A flexible manufacturing cell. (Photo copyright 2000, courtesy FANUC Robotics North America, Inc., reprinted with permission.)

have been contributing to this development since it began. However, there seems to be recognition that the fully automatic factory (sometimes referred to in the popular media as the "lights-out factory") is just as far off as the fully autonomous robot in any other application — notwithstanding the fact that a few demonstrations of combined automatic operations have succeeded in running for days at a time without need of human intervention.

Organization of the whole factory in space and time includes both flow of materials (factory layout) and flow of information (scheduling). Factory layout and scheduling are probably the most difficult ongoing problem for industrial engineers. They are discussed further in the section titled "Optimization" in Chapter 3.

The Tasks for Humans

Parts are found to be defective. Tools wear. Machines fail and have to be repaired and maintained. Manufacturing objectives change, and the system must be reprogrammed from time to time. Realistically, both humans and computers are required.

Important human decision functions regarding the flexible manufacturing system include scheduling of production changes, constraints of limited resources, monitoring of system flows and outputs, scheduling of maintenance, and intervention in the case of failure. Planning and scheduling are essential to reduce item flow times, reduce work-in-process inventory, increase machine utilization, guarantee product quality, and meet promised due dates. There are physical constraints on the rates and accuracies of materials handling and transfer (with some important trade-offs) and constraints on parts feed and cutting or processing speed (more trade-offs). There are economic constraints on dollars as well as time, safety, and accuracy (quality) constraints.

Mathematical models have addressed only a few variables and issues, and artificial intelligence and advanced control can accommodate only those factors that are well defined. The numbers of variables, possible configurations, and salient qualitative factors in the flexible manufacturing system are immense. Therefore the human operator must plan and manage. Carefully designed displays and controls and computer-based simulation, prediction, and decision support are essential.

Some human operators in the modern manufacturing plant make their way among the machines, inspecting parts, observing displays, and modifying control settings or keying in commands, most of it through computer-mediated control panels adjacent to the various machines. Some human operators do their work from a centralized control room, which makes essentially all their observations through computer-mediated displays and controls. In both cases the operators may be said to be *supervisory controllers*, and the machines may be said generically to be *telerobots* (more on this in Chapter 6). The etymology of *manufacture* — *manus* (hand) plus *factus* (past participle of *facere*, to make) — is no longer applicable.

Hospital Systems

Computer-based automation is finding application in all branches of health care. The hospital operating room and the intensive care unit are coming to look more like process plant control rooms, with many panels of displays and controls mediated by computers. For example, the anesthesiologist is really a process supervisory controller; the process is the patient, and the anesthesiologist plans a drug treatment, programs the machinery in the operating room to dispense the correct anesthetics in the correct amounts at the correct time, monitors the patient's vital signs, and intervenes as necessary to change the treatment or directly handle the patient in other ways (or advise the surgeon to do the same; Cook & Woods, 1996; Xiao, Milgram, & Doyle, 1997).

Most hospitals recently began using computer systems for physicians to pass instructions to nurses, to order medications, and so on. This enables much faster and more reliable communications to laboratories and to pharmacies. Such systems avoid problems with illegible physician handwriting and enable automatic checking with patient medical data so that conflicts between medications, overdoses for infants (based on weight), wrong site surgery, and other medical errors can be caught before it is too late. Such systems can also be used to encourage more complete information to be passed when patients are being handed off from one physician to another or from one nurse to another. However, poor typing skills and unfamiliarity with computer interaction remain as problems with hospital staff, though this is improving with newer, younger employees.

Programmed robots under supervisory control by the orthopedic surgeon are now being used to drill specially shaped holes in bones during operations to insert artificial joints. The holes thus drilled can ensure much tighter and cleaner fits than manually drilled holes.

Medical Monitoring in the Home with Communication to the Hospital

Miniaturization of sensors and radio communication devices has made possible a whole new prospect for health monitoring of patients with chronic problems. These advances enable automatic monitoring of patients in the hospital or chronic care facility, with alarms sounded in the nurses' station when variables exceed their set thresholds. There are exciting new prospects for instruments that are operable by patients living normally in their own homes (or while working, shopping, etc.) which can signal the appropriate medical personnel by means of the cellular telephone or personal digital assistant (PDA) communication infrastructure.

In the next decade a significant increase is expected in medical technology that moves into the home and allows persons with chronic problems of many kinds (or their loved ones or home health aides) to monitor their condition on an hourly or daily basis. Home-based computers will make preliminary diagnoses and, if there is cause for concern, will relay data over high-bandwidth telephone links to professional diagnosticians in the hospital who can interpret changes relative to the patient's computerized medical record. Treatment advice can be given, medical appointments can be made as necessary, or, under critical conditions, an ambulance can be called. Figure 2.10 illustrates the logic of how home monitoring can effect a kind of triage to relieve the patient of anxiety, reduce wasted time on hospital visits, and relieve the load on hospital emergency facilities.

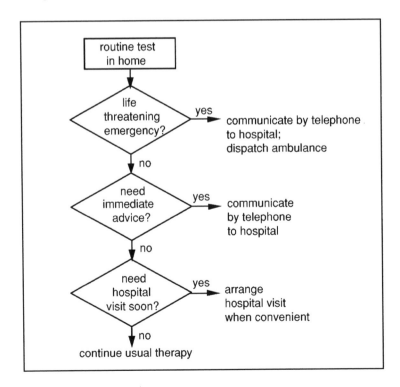

Figure 2.10 Triage logic for home monitoring and telecommunication with hospital.

Minimally Invasive Surgery

Minimally invasive surgery refers to operations performed on internal body structures through sealed puncture devices (trochars) by means of miniature video cameras or fiber optics (endoscopes) to view inside and special tweezers, scissors, knives, staplers, and so forth that are handled from outside but inserted through the trochars to manipulate inside.

A few endoscopic operations are now common, such as cholecystectomy (gall bladder removal) and repair of torn cartilage in knee joints, but there is much excitement about a broad range of new surgical procedures. There is little doubt that automation will soon be added to what is now essentially manually controlled endoscopic surgery. This is already happening in brain surgery, where computer-based images from prior magnetic resonance imaging scans are superposed on the skull so that the surgeon knows where to open the skull and insert instruments.

Telemedicine and Telesurgery

Telemedicine refers to the delivery of health care over communication channels, and that can mean two-way video, audio, and data channels connecting the patient and a nurse or paramedic accompanying the patient in remote areas (in space or on ships at sea) to the medical specialist at a big city hospital. Experiments with telemedicine have been promising, but the medical culture has been somewhat slow to adopt telemedicine to any great extent.

Supervisory control in this case often means that the more experienced but remote specialist supervises the nurse or paramedic who is local to the patient in doing examinations and performing therapeutic tasks (Ottensmeyer, Hu, Thompson, Ren, & Sheridan, 2000). The U.S. Army is even experimenting with telesurgery, combining teleoperation over a communication channel with endoscopy to better treat wounded soldiers on the battlefield. The idea is to stop the bleeding rapidly so that the soldiers can be safely returned to field hospitals.

Figure 2.11 diagrams a telesurgery system from a research perspective in which the aim is to explore the effects of communication constraints on system performance. The four blocks designate information "filters," which constrain the remote surgeon from getting or sending perfect information and executing perfect control. G_{Ae} is the constraint on instructions to or effects on the human assistant. G_{As} constrains the sensory feedback from the assistant and the video scene. G_{Te} constrains the commands to or effects on the teleoperator. G_{Ts} constrains the sensory feedback from the teleoperator. It is not yet understood what constitutes these constraints and their relative effects, whether in telesurgery or in other tasks in which

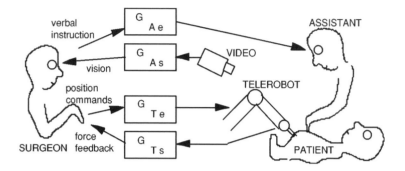

Figure 2.11 A telesurgery system showing the four communication "filters."

humans communicate remotely with combinations of automation and other people. The four filters merely provide a way of posing the problem.

Command and Control in Defense and Civilian Emergencies

Another major type of human-machine system in which automation is playing an increasing role is the *command and control system*. These are thought of primarily with respect to large-scale military operations, but they can also be important in civilian emergencies involving police, fire, weather, and terrorist attacks. In one sense every automation system is a command and control system, but here I am referring to large-scale, multi-person, geographically distributed command and control systems. A popular alternative phrase is *command, communication, control, and information* (C^3I) *system*. In these applications the present concern is primarily with communication of information, not the physical hardware per se.

In its simplest (and perhaps least realistic) form, command and control is hierarchical, with a person on top giving orders down a pyramid of lower-status individuals and getting feedback sent back up. A somewhat more realistic (and chaotic) representation might be that shown in Figure 2.12, in which sensors, decision elements, and actuators, each of which can be either human or artificial, are interconnected in complex causal relations. To make matters more complex, there can be time delays and noise imposed in the links, possibly resulting in instability or confusion in decision because of conflicting data from different sources. This complexity contributes to what military scholars call "the fog of war."

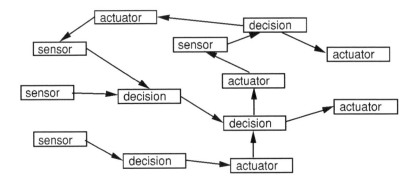

Figure 2.12 The complexity of real command and control in ever more complex systems.

Office Systems

Office automation has taken many forms. The printing press and type-writer were the beginning, then the mechanical calculator, next the multi-function pocket electronic calculator, and recently the personal digital assistant or PDA. Speech recognition systems are now common and are becoming substitutes for dictation and human secretaries. E-mail makes interpersonal communication faster, cheaper, and more convenient. Copper wire, then coaxial cable, and now optical fiber and satellite networks permit computers to talk to each other, accessing data, text, and images from each other at high bandwidth. Video teleconferencing, available for many years at a high price, has become readily available at the Internet terminal.

Business software simplifies purchasing, shipping, accounting, tax filing, and every other function for which, heretofore, hard-copy forms had to be filled out and mailed from one place to another. Publishing in either hard copy or electronic form has been greatly simplified. Engineering and architectural design, music composition, library research, and other creative activities are conducted on computers. Industrial training programs as well as university courses are put on the Web. Internet tele-communication and marketing is surely the fastest-growing sector of the economy.

Unfortunately, predictions that less paper would be used have not come true, because much more text is typed and more hard copies are printed, saved temporarily, and thrown out. More communications go to more people faster, with the result that more time is spent reading e-mail that is of no interest but that must be dealt with at some level. Advertising imposes itself on most Web interactions. There are threats to privacy as well as pornography and other affronts to cultural values. In short, the microcomputer in the business, government, or home office has thrust people into engagements that make their working day far different from (and, arguably, not always better than) that only a few decades ago. One major effect on people has been that coworkers do not have to work in close physical proximity. They can work at home, or at least on more flexible schedules, with the computer as the buffer/mediator. It also means that small and large firms can function interactively with one another. Outsourcing of services as well as goods is much easier and more extensive than before. These factors, together with supplier divestiture from the parent firm and continued increase in communication demand, have produced during the period 1960–2000 a threefold growth in U.S. telephone communications and a threefold decline in cost per call. Computer-supported cooperative work, in which people work together over electronic communication and computer intermediaries, is a recognized discipline among industrial managers (Greenberg, 1991).

Education

The first teaching machines appeared almost 50 years ago in B. F. Skinner's laboratory at Harvard University (I was a participant!). They were cardboard disks on turntables that posed questions and told the users whether or not their answer (indicated by a stylus that made electrical contact through the holes in the disk) was correct. They have come a long way since then, but formal education is still mostly a matter of students assembling in classrooms and writing notes on paper while the teacher writes on a chalkboard.

Web-based pedagogical text and images are now common in universities and are becoming so in high schools. Some, but not much yet, is interactive. As bandwidths into classrooms, dormitories, and homes improve, there will be many more opportunities for feedback, detailed and animated graphics, and so-called virtual laboratories where the students manipulate the configuration or adjust the parameters of virtual physics experiments, chemical reactions, and mechanical designs. Hands-on, real-time interaction is key to learning, but whether or not that interaction must be with real physical objects and real people is a question yet to be resolved.

Libraries of the future are already accreting on the Internet, although access to them is still limited to the minority who have access to computers and know how to use them. One serious problem is reading from computer screens. Most users still prefer to print out documents to read them. Computer screen resolution is gradually improving, but there remains evidence that eyestrain and headaches result from images that are slightly unstable at the pixel level. Traditional librarians have serious questions about what the computer is doing to the book. The jury is still out.

The "Smart" Home

Automation has been entering the home in many ways. As mentioned in the last chapter, toilet tank water level control and temperature control for home heating (in the form of the simple on-off thermostat) were among the earliest forms of home automation. Air conditioning was added to heating, and then humidity control appeared. Now one can program different parts of the house to maintain different temperatures at different times of the day. Future home heating may allow the homeowner to store heat as electrical power locally so as to reduce the daily peak demand for electricity and allow for more economical power generation.

Computer control of home appliances has gradually become more sophisticated. Clothes washers and dryers and dishwashers have more program settings. They have also become more self-adaptive to conditions: Clothes washers will shut off the spin cycle if the balance is not sufficient, dryers can adapt the drying time to the weight or moisture content of the clothing, dishwashers automatically cycle — and all this is taken for granted by the user. Stoves, microwave ovens, and toasters are also programmable with respect to time and temperature and are adaptive with respect to temperature sensors implantable in the item being cooked. Electric irons have many more features (steam, for example) and automatically shut off when not in the correct position.

New software systems are becoming available to track, reorder, and even deliver groceries and other supplies for the home, including medicines. Electronic tagging of products will enable retailers to track and replenish inventory, customers to have their cooking time or prescription conflicts monitored by their home computers, and suppliers to track customer preferences and target advertising.

Home security systems are a growing market, although current manifestations are not always dependable and have a tendency toward false alarms (a problem to be discussed in later sections). It is not always easy for them to discriminate between malevolent intruders and pets, friendly visitors, or family members who forget to reset the system. Neighbors and police too often experience alarms that are regarded as "crying wolf."

Sleeping children can be monitored from other rooms or even from down the street with radio transmission. Slow-frame video may soon be added.

Home entertainment has seen many recent changes. Live broadcasts evolved from radio to TV to high-density TV, with tens or hundreds of channels to select from. Many people routinely record the news or other programs and play them back at more convenient times. Recorded music evolved from phonograph recordings to tape cassettes to compact discs. Music can now be piped anywhere in the home, provided through portable playback devices, or downloaded from the Internet. The quality of musical recording and playback has reached the point that listening to a concert in the home is a more faithful rendition than listening from the rear of the concert hall.

Computers have invaded children's lives in ways that are astounding and disturbing. Beyond the fact that the average child in a developed country spends a significant portion of time in front of the TV, one sees myriad, increasingly sophisticated computer-interactive games and toys of all kinds. It is probably true that the newer generations are less fearful of computers because of this experience as children. It may also be true, however, that they are cheating themselves out of traditional social and creative activities that have greater value in learning to cope with the world in which they will have to live.

"Built-In" Automation and Evolution of Tasks

Few, if any, of the devices just described required technological break-throughs, although because of huge markets, much technological development has taken place. Human *usability testing* has become prevalent, based on questionnaires, hidden cameras, focus groups, and similar techniques. Use of this technology is market driven, so the question is, what will people be willing to pay for? The question is driven partly by the unsubtle influence of advertising and only partly by need or usefulness.

Consider the case of household robots. For many years the futuristic literature and the media in general have suggested that robots would be available well before the year 2000 to perform household chores. People imagined humanoid robots of the R2D2 variety scurrying around the house sweeping, dusting, and generally picking up. That has not happened, and I am skeptical that it will. Many household chores are complex sensory-motor skill tasks that are slightly different each time they are done, and programming general-purpose telerobots to accommodate to these subtle differences would be difficult and time consuming. What is happening is that a variety of special-purpose automation features are being built into various systems, as described in the preceding paragraphs, which also implies redesign of the household chores themselves. There will still be home chores for humans to do, but the tasks will continue to change.

Automation on the Person

One of the more dramatic implementations of automation, defined broadly, is on the person. What automatic devices are being carried by, worn on, or implanted within the body? The pocket watch and later the wristwatch were perhaps the first items of on-person automation. More recently, pocket and wristwatch calculators and organizers appeared. Following legislative initiatives in several countries to limit monopolies and open markets in telecommunication, cellular phones came into widespread use. Now communication, calculation, and memory-access functions are being combined with cellular phone functions into integrated handheld or wrist-worn devices. The supermarket shopper who once appeared to be talking to the vegetables is now easily recognized as using a head-mounted cell phone, thus allowing her hands to be free to select groceries.

Hearing aids are highly refined pieces of on-body automation, becoming smaller and more adaptable to alter spectral shape and loudness to suit the wearer's particular hearing losses. Eyeglass-based computer displays have recently emerged. One such device allows the user to view the environment with one eye while presenting stored text or graphics to the other or to otherwise superimpose the two visual fields. These devices can be used, for example, by maintenance personnel who want to look at an actual machine and review instructions for disassembly or testing at the same time.

Plastic credit/debit cards, which heretofore have been passive, enabling only identification information to be read by bar code or similar readers, now contain chips that allow accounts to be debited or credited. Devices are being developed whereby two people can selectively exchange information about themselves from their "smart cards."

Space suit engineering has demonstrated that heating and air conditioning can be put directly on the body. If energy costs were to increase, heating bodies in a similar manner might be a viable alternative to heating buildings.

Today technologists are experimenting with a great variety of applications of wearable computers: chips built into clothing, or what one of my colleagues calls "underware." Because Internet connectability will be much more ubiquitous than before — from practically anywhere you happen to be (Negroponte, 1995) — many more functions can be imagined. When combined with GPS and video sensors, the Internet might possibly keep track of where a person (a child or a criminal on probation, for example) is located and what he or she is doing and call on nearby human assistance as needed.

The field of rehabilitation engineering has gradually seen prosthetic (substitution for missing limb) and orthotic (bracing for existing but nonfunctional limbs) adopt automation. Upper-extremity devices were among the earliest to have automatic grasping of an appropriate force and automatic finger bending depending on where an object is grasped. Lower-extremity devices now have electric brakes coupled with logic to make the gait smooth and natural. Cardiac pacemakers and their implanted electrodes are gradually improving. Electrodes implanted in the occipital cortex are being driven experimentally by computers to provide primitive sight for the blind. Other computer-driven implanted devices are being used to regulate the delivery of critical drugs. Stemming from developments in medical monitoring for astronauts and foot soldiers, a variety of new wearable or implantable sensing and local processing devices are being considered for monitoring patients with a variety of chronic diseases.

PART II: DESIGN OF HUMAN-AUTOMATION SYSTEMS

Chapter 3

THE ANALYSIS AND DESIGN PROCESS
The Role of the Human Factors Professional

The purpose of design, from the point of view of the human factors professional, is to achieve a proper relationship between the behavior of the hardware and software technology and the behavior of the human user. Many engineers have been conditioned to think that design is always design of a physical thing, so design of a relationship to a human user is a difficult concept.

Human factors professionals are trained to start from the viewpoint of the human user (and from the fact that designer "common sense" may be flawed). Their training is varied, as some start from an engineering base, some from an experimental psychology or physiology base, and some from a base of medicine, dentistry, or another applied human science. The common viewpoint is an orientation of experimental science, an appreciation for the variability among people and the variability of behavior within the same person. They also understand the use of statistics to make rational inferences from observations, and they have the ability to apply mathematical modeling where appropriate. There are boards of certification for human factors professionals, but currently there is no consensus on whether certification is an important pedigree.

The human factors professional will usually insist on performance tests using either the real environment (field tests) or some type of simulation (if the system under consideration is not yet in the real world). He or she will typically ask many questions that may seem antagonistic to the design engineer, who often assumes the protagonist role for any new design; nevertheless, taking an initially neutral stance toward the proposed benefits of any new design is a sacred obligation of the human factors professional. For these reasons it is important to bring the human factors viewpoint into consideration early in the design process, not later, after the design has been frozen and there is little real opportunity to correct any shortcomings that are found.

Human factors professionals should not be segregated from the design engineers. Human factors types may feel more comfortable maintaining their own identity in their separate ivory towers, surrounded by their own laboratories and their own jargon. However, they need to expose themselves to and learn the jargon and the ways of thinking of the design engineers, bringing the user representatives along into that same environment, as appropriate. Only then will they win the respect of the engineers and see their advice implemented. Simply publishing research results in learned journals and posting human factors guidelines (much as Martin Luther posted his ultimatums on the church door) often leads to being ignored by the engineers.

Accident data banks are available and can be very helpful. Anonymous self-reports of aircraft near-misses by pilots have resulted in a widely published NASA/FAA database. The U.S. Department of Transportation has, over many years, compiled an extensive database of fatal auto accidents. Confidential accident and incident data collected within industry, NASA, FAA, the Department of Transportation, Department of Defense, Nuclear Regulatory Commission, and elsewhere can easily be sanitized and made available for general use. Although every accident or near-accident is unique in some respect, with enough data, some powerful and useful inferences can be drawn.

Human-Machine Task Analysis: Goals and Constraints

Task analysis is commonly regarded as the first step in the human-system design process. Task analysis requires the breakdown of overall tasks into their elements and the specification of how these elements relate to one another in space, time, and functional relation. Who (including which person, if there are several, or which machine) does what, when, and how? By *task* one can mean the complete performance of a given procedure, the totality of activities to design and/or build a given thing, to monitor and/or control a given system, or to diagnose or solve a given problem. Alternatively, *task* can mean a small subelement, such as a particular movement or measurement. Sometimes *mission* is used to connote the full task or end state to be achieved, and *subtask* is used to refer to some component. Terminology for such breakdown of tasks is of no particular concern here except to note that some hierarchical breakdown is usually helpful or even necessary.

When a system is being automated, or existing automation is being improved, it is particularly important to rethink the task analysis. This is because automation can change the nature of demands and responsibilities, often in ways that were unintended or unanticipated (Boy, 1998; Parasuraman, Sheridan, & Wickens, 2000; Parasuraman & Riley, 1997; Wiener & Curry, 1980; Woods, 1996).

The starting point is usually a set of implicit task constraints, and task analysis is really a matter of articulating these constraints and making them visible, being very clear about what are independent and what are dependent variables.

Although many techniques are used for task analysis (Kirwan & Ainsworth, 1992), it is still very much an art. Task analyses end up being verbal statements of missions with qualifications, mathematical equations, block diagrams defining elements with arrows indicating what influences what, lists of variables with ranges and/or statistical properties, timelines, flow charts, and so forth.

When performing a task analysis, it is all too easy to assume more constraints than are really there or are necessary. For example, there is a tendency to fall into the trap of writing down the series of steps a human takes to look at particular already existing displays and operate particular already existing controls, the very displays and controls that constitute the present interface one seeks to improve or to automate.

The correct way to analyze the task is to specify the information required, the decisions to be made, the control actions to be taken (at the level of the controlled process), and the criterion for satisfactory completion of that step (Figure 3.1). This should be independent of the particular human or machine means to achieve those steps — that is, the particular displays or controls that already are, or even those that might be, used for task implementation.

Following the ecological perspective (Vicente, 1999; see the section on ecological displays in Chapter 5 of this book) and the idea of *joint cognition* (Dowell & Long, 1998; Hollnagel & Woods, 1983; Rasmussen, Pejtersen, & Goodstein, 1994; Woods & Roth, 1995), there is a call for postponing the traditional task analysis until after the work domain constraints have been thoroughly spelled out. Studying and specifying work constraints, it seems to me, is an obvious part of defining any problem, and if any task analysis avoids this, it is to be faulted.

Task step	Operator or machine identification	Information required	Decision(s) to be made	Control action required	Criterion of satisfactory completion

Figure 3.1 An example format for task analysis.

Human interaction with the environment is a closed-loop process: The human observes the environment and in turn affects the environment. Causation goes both ways, as depicted in Figure 3.2. Therefore it is important to think of the constraints that operate in both directions: all the factors that constrain the human from observing the environment fully and perfectly, and all the factors that constrain the human from action upon the environment exactly as he or she might wish. In Figure 3.2 these constraints are subsumed under the rubric *filters*. It is important to keep track of which factors act as filters or constraints in the human observation (affective) direction and which act in the human motor action (effective) direction.

It is not easy to do the task analysis without implying some *function allocation* — what functions should be allocated to humans and what functions to machines — which is really part of system synthesis. Nevertheless, the effort must be made to determine what is necessarily a given and what may be left as degrees of freedom for synthesis.

Human-Machine Function Allocation

Function allocation requires the designer to assign various functions (tasks, jobs) that need to be done (as specified in the task analysis) to resources — that is, to particular humans, instruments, or machine agents capable of doing those tasks. Alternatively, function allocation can be thought of as the allocation of such resources to tasks, which is equivalent.

Many tasks must be done by a human because it is not known how to automate them. Other tasks are done by humans because humans like to do them. Still others are done by humans because the tasks are trivial or are embedded in either of the other two types of task, and therefore it is inconvenient to devise some automation to do them. To the extent that a task now done by a human is boring, fatiguing, or hazardous, it should be automated if feasible and convenient to do so. Otherwise, when a task can

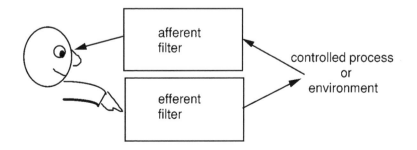

Figure 3.2 Afferent and efferent filters between human and environment.

be done by either human or machine, it is worth considering whether human and machine might work together and complement each other.

Currently, although many accepted techniques exist for doing task analysis, there is no commonly accepted means to perform function allocation (and especially to "optimize" this allocation, a bit of semantic nonsense that one sometimes sees in print). The reasons are several. First, although tasks/functions (the "whats" and the "hows") may indeed be broken into pieces, those pieces are seldom independent of one another, and the task components may interact in different ways depending on the resources chosen for doing them. Second, the human and computer can interact in an infinite number of ways, resulting in an infinite spectrum of allocation possibilities from which to choose. Third, criteria for judging the suitability of various human-machine mixes are usually difficult to quantify and often implicit. Although I suggest some things to consider when allocating tasks to people and machines (especially machines with so-called intelligent capabilities), I make no attempt to offer a general procedure for synthesizing the task allocation.

The Fitts MABA-MABA List

Fitts (1951) proposed a list of what "men are better at" and what "machines are better at" (MABA-MABA; see Table 3.1). This is sometimes called the Fitts MABA-MABA list or simply the Fitts list. It is often considered to be the first well-known basis for task allocation. An expanded version appears in NUREG-0700 (U.S. Nuclear Regulatory Commission, 1981).

Jordan (1963) was concerned that the Fitts list might be used by people who assume that the goal is to compare people with machines and then to decide which is best for each function or task element. He quoted Craik (1947a, 1947b), who had pointed out that to the extent that the human is understood as a machine, it can be known how to replace a human with a

TABLE 3.1
The Fitts MABA-MABA List

Men Are Better At	Machines Are Better At
Detecting small amounts of visual, auditory, or chemical energy	Responding quickly to control signals
Perceiving patterns of light or sound	Applying great force smoothly and precisely
Improvising and using flexible procedures	Storing information briefly, erasing it completely
Storing information for long periods of time and recalling appropriate parts	Reasoning deductively
Reasoning inductively	
Exercising judgment	

machine. Early studies by Birmingham and Taylor (1954) revealed that in simple manual control loops, performance can be improved by quickening, wherein visual feedback signals are biased by derivatives of those signals, thereby adding artificial anticipation (what the control engineer would call *proportional-plus-derivative control*) and saving the human operator the trouble of performing this computation cognitively. Birmingham and Taylor concluded that "man is best when doing least" in this case. Jordan suggested that this is a perfect example of Craik's tenet. He also quoted Einstein and Infeld (1942), who discussed the development and then the demise of the concept of ether in physics and how, when empirical facts do not agree with accepted concepts, it is time to throw out the accepted concepts (but retain the empirical facts). Jordan's point was that the idea of comparing the human with the machine should be thrown out but that the facts about what people do best and what machines do best should be retained. Jordan's idea of what should be espoused, and the main point of retaining the Fitts list, is that people and machines are complementary.

Also in keeping with Craik's (1947a, 1947b) tenet, Jones and Jasek (1997), Woods (1996), and Hoc (2000) emphasized that the goal is human-machine cooperation — not the allocation of separate functions to separate entities — and this includes cooperation in planning and action. However, exactly what it means to cooperate is still to be worked out in terms useful for human factors engineering.

It might seem that a straightforward procedure for doing task allocation would be to write down all the salient mixes of allocation and then write down all the applicable criteria. Following this, one might rank order all combinations of allocation mix and criteria, thus determining a rank-order score. Alternatively, one could weight the relative importance of each criterion, rate each mix by each criterion, multiply by the weight, and add up the scores for each allocation mix. However, there can be difficulties with any such direct method — hidden assumptions, unanticipated criteria situations, nonindependence of tasks, nonindependence of criteria, nonlinearities that invalidate simple multiplication of weight by rating and addition of products — in particular, the fact that a very large number of possible interactions between human and computer compete for consideration; it is not a simple matter of "human versus computer," a point that is emphasized repeatedly in this book.

Price (1985) asserted that in order to make use of the Fitts MABA-MABA list, one needs data that are context dependent, but these data are mostly unavailable. The difficulty in acquiring these data is exacerbated by the fact that the status of the machine is not static; the capabilities of machines to perform intelligent acts such as automation and decision support are ever improving. However, Price claimed, automation can "starve cognition" if the human is not kept in sufficient communication with what the automation is doing or intending. He seems to have agreed with Jordan (1963) when he pointed out that human performance and machine performance are not a zero-sum game, implying that the combination can be much better than either by itself. Kantowitz and Sorkin (1987), Price (1990), and Volume 52 of the *International Journal of Human-Computer Studies* (2000) provide reviews of the literature in task allocation.

One can make a distinction between tasks as goals to be achieved and functions as means to achieve the goals, what the Fitts list specifies as best for each of human and machine. If two functions within a task are closely related and interdependent, it may be inadvisable to allocate one to the machine and one to the human. For example, for air traffic control, Bisseret (1970) recommended against systematically assigning the conflict detection function to the machine and the resolution function to the human, suggesting rather that the pair of functions be assigned to either human or machine depending on the type of conflict. In other words, assignment should be by task rather than by function (Vanderhaegen, Crevits, Debernard, & Millot, 1994).

According to Rasmussen (2000), human-automation design should not be based on normative work procedures but, rather, should allow a "resource envelope" within which the operator can work freely and still maintain the support of the automation.

The Fitts list generalizations remain a useful starting point for allocation of functions between human and computer. No other allocation model has replaced it in terms of simplicity and understandability. However, it is a qualitative statement subject to interpretation. As computers become "smarter," they creep closer to bettering human capabilities. One can say for sure, based on lots of evidence, that human memory prefers large chunks with interconnected or associated elements, prefers relatively complete pictures and patterns, and tends to be less good at details. This empirical fact strongly suggests that insofar as it is feasible and practical, the human should be left to deal with the "big picture" while the computer copes with the details.

Whether function allocation will ever become more a science than an art is debatable. In fact, the issue has sparked much recent debate. Fuld (2000) advocated a bottom-up rather than a top-down approach, believing that function allocation is a useful theory but not a practical method. In Sheridan (2000), I asserted that the human-automation mix will necessarily and unstoppably evolve because of forces of technology and culture, but that does not excuse the designer from doing his or her best, especially when human values are at stake.

The public (as well as, unfortunately, too many political and industry decision makers) has been slow to realize that function/task allocation does not usually mean allocation of a whole task to either human or machine, exclusive of the other. For example, in the space program it has been common to consider that a task must be done by either an astronaut or a robot; that if a spacecraft is staffed, then astronauts must do almost everything; and that if a spacecraft is unstaffed, every task must be automated. In fact, on occupied spacecraft many functions are automatic, and on unoccupied spacecraft many functions are performed by human remote control from the ground. In the next section I show that different stages of a task can be separately automated to greater and lesser degrees.

Allocation at Four Stages and in Differing Degrees

When considering how far to go with automation and decision aiding, the system designer should ask what parts of the system cannot be automated because it is not known how to automate them, what parts humans like to do, what parts are trivial for humans and inconvenient to automate, and what parts are boring, fatiguing, or hazardous and thus should be automated if possible. Also, what degree of unfamiliarity can humans cope with, and what degree of unfamiliarity can be accommodated by automation using the best available predictive models? Surely automation need not apply uniformly to entire tasks; rather, it can be of more benefit to some parts than to others.

A taxonomy that applies to most complex human-machine systems is the sequence of operations shown in Figure 3.3: acquire information,

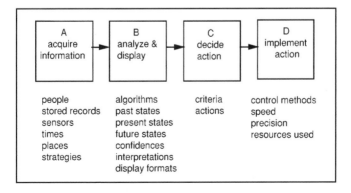

Figure 3.3 Four stages of a complex human-machine task.

analyze and display, decide action, and implement action. Below each box in this figure are noted some typical variables relevant to that stage.

Table 3.2 presents an eight-level scale of degrees of automation, a simplification of an earlier scale (Sheridan & Verplank, 1978). To make it simple, the scale folds together several relevant dimensions, such as

- the degree of specificity required of the human for inputting requests to the machine,
- the degree of specificity with which the machine communicates decision alternatives or action recommendations to the human;
- the degree to which the human has responsibility for initiating implementation of action, and
- the timing and detail of feedback to the human after machine action is taken.

It is evident that these many dimensions of automation can be ordered in various ways on multidimensional as well as one-dimensional scales. The main point is to have some way of considering many options that are more or less ordered and then to decide the degree of automation at each stage. There is no reason to force each stage to the same level of automation.

Figure 3.4 illustrates how different kinds of tasks (represented by the three types of lines) call for different levels of automation at different stages. The numbers correspond to the eight-point scale, Level 1 representing the *least automation* and Level 8 the *most automation.* The circles might represent, for example, the task of holding an election of officers for an organization (Sheridan, 1998). At the *acquire* stage, data are collected manually from eligible individuals, although e-mail (Level 2 automation) might have been used to make suggestions to solicitors on how to operate. The results are then *analyzed* by computer, and the winners are *decided* automatically. The transfer of power is then *implemented* with only a modest amount of computer-stored procedural advice.

TABLE 3.2
A Scale of Degrees of Automation

1.	The computer offers no assistance; the human must do it all.
2.	The computer suggests alternative ways to do the task.
3.	The computer selects one way to do the task and
4.	executes that suggestion if the human approves, or
5.	allows the human a restricted time to veto before automatic execution, or
6.	executes automatically, then necessarily informs the human, or
7.	executes automatically, then informs the human only if asked.
8.	The computer selects the method, executes the task, and ignores the human.

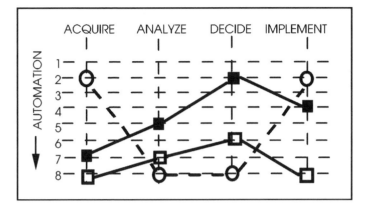

Figure 3.4 Examples of different levels of automation at different task stages.

The black squares might represent advice recently given by a U.S. National Research Council (Wickens, Mavor, et al., 1998) committee advising on the proper amount of automation for new civilian ATC systems. After much debate, the committee decided that acquisition and analysis could and should be highly automated — in fact, they already are (radar, weather, schedule information, etc.) — but that decision making, except for certain decision aids now under development, should be done by human air traffic controllers. Implementation is in the hands of the pilots, which in turn is largely turned over to autopilots and the other parts of the Flight Management System.

The open squares might represent a typical manufacturing robotics task. A computer vision system acquires all the data. This is analyzed in a computer where the analysis results are available to a human supervisor should he or she need to check the data. The analysis is passed on to a computer decision algorithm, and the decision results are displayed for the operator. The decision is implemented by a robot in fully automatic fashion.

As part of function allocation, Wei, Macwan, and Wierenga (1998) suggested a model for the appropriate degree of automation of different tasks, based on a task's effect on system performance and its demand on the operator relative to other tasks. Experimentally they showed that as automation was increased, performance did not improve beyond a certain level.

Human-Computer Trading and Sharing

Human and computer can either pass control alternately back and forth (called *trading*) or can function simultaneously (called *sharing* the task).

Both trading and sharing may occur as parts of the same overall task. Figure 3.5 conveys the idea: The connections in the top part of the figure alternate between human and computer, whereas in the bottom part they remain fixed in parallel operation.

Trading is necessary when the human must do some of the task and then turn it over to the computer to do other parts. For example, developing software involves both programming and testing alternately. Trading is often recommended when the human or machine performs a task and the other does a check. It can also be useful if the one that normally does a task is busy and the other can step in on a temporary basis to settle urgent matters or "buy time."

Sharing is useful when human and machine can do different portions of the same task — for instance, if the human controls in some degrees of freedom while the machine simultaneously controls in others. This may be necessary in some real-time control or monitoring tasks, or it may be a matter of making the best use of the human and machine resources.

Adaptive Automation

A number of researchers have pointed out the need for *adaptive automation*, meaning that the allocation of functions between human and automation is not fixed but varies with time depending on the human's momentary workload as well as the current context (Hancock & Scallen, 1996; Kaber & Riley, 1999; Moray, Inagaki, & Itoh, 2000; Parasuraman, Mouloua, & Molloy, 1996; Rouse, 1988). A change in function allocation could also be at the request of the human (asking for help) or at the

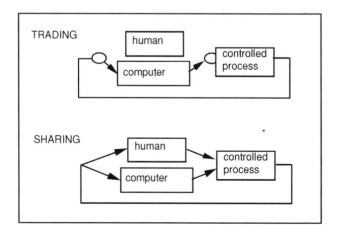

Figure 3.5 Trading and sharing.

insistence of the automation (telling the human the job is not getting done, or simply taking over when the human is napping or otherwise preoccupied).

Large System Organization

In the preceding sections I have discussed task analysis and function allocation at the level of the workplace for a single human operator or a small team. When many machines and people make up an operating system, the task analysis and function allocation must be viewed at a higher level of system organization (as well as at the lower level) with respect to both spatial allocation (building architecture, traffic system structure, etc.) and temporal dynamics.

A good example is the organization of a modern factory. Modern manufacturing is driven by cost, and major determinants of cost are wastage of human operator time and wastage of parts inventory. For this reason, the organization in space (factory layout) and the organization in time (work scheduling) are critical. One normally thinks of a factory as a system that processes raw materials to make parts, which are then assembled to make products, which are then shipped to the customer (which might itself be a conventional wholesaler, an Internet wholesaler, a retail store, or an individual). This material flow is shown from left to right in Figure 3.6. What has often been neglected in the past is the information (requirements) flow (right to left in the figure). Three feedback loops are operative: between raw materials and material processing operations (machining, forming, welding, etc.); between material processing and parts assembly, and between parts assembly and customer demand. Each information requirement provides a set point to the control loop to its left.

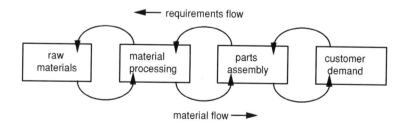

Figure 3.6 Basis for organizing a factory in space and time.

Optimization

The system designer would like to have the goal(s) or objectives of the system specified quantitatively, stated as relative goodness of performance (a scalar) as a function of system inputs and outputs and resources used. This is called an *objective function* or *utility function* (Keeney & Raiffa, 1976). (Objective functions and the trade-off problem are discussed formally in Appendix A.) This is also true of given physical and economic constraints, such as size, force, time, and money.

The controlled process — that is, the relations between independent and dependent variables of the physical thing being controlled (the aircraft, automobile, nuclear reactor, etc.) — will likely already have been modeled, or at least approximated, mathematically, for formal closed-loop control requires at least an estimated process model. The automatic controller can also be modeled if it is simple enough; for instance, if it is a sum of differently weighted proportional, time-derivative, and time-integral functions of the control error (system state deviation from what is wanted), such as make up linear control theory (see Appendix A).

In the engineer's ideal world, one could write explicit equations for (a) the controlled process, (b) the constraints (physical, economic, and human), and (c) the objective function. Then, by simultaneous solution of these equations, one might maximize "goodness," and thus an optimal control strategy would be determined. This, in effect, is what is done in optimal control theory, linear and dynamic programming, and myriad other mathematical techniques for optimization. The reason these techniques are seldom applied to real-world systems is that the assumptions cannot be fulfilled. One simply does not have the necessary information in explicit mathematical form.

Given that real-world optimization of purely physical automation systems is sometimes tractable (because a mathematical objective function is available to specify what is better and what is worse), what about human-automation systems? Can they be optimized? Here it has to be asserted that optimization in the formal sense is simply out of the question, because the cognitive contribution of the human operator can hardly be put in mathematical form. Thus claims or stated intentions of optimization of a human-machine system are meaningless, at least in the formal sense that optimization is a unique state.

What is much more likely is gradual improvement of systems relative to (sometimes changing) criteria — attributable to iterative redesign and operator/user experience over a long period of evolution (Sheridan, 2000).

Handbooks, Guidelines, and Standards

Over the half-century that human factors engineering has existed, various handbooks on that topic have been published, probably the best known of which are Van Cott and Kinkade (1972), Boff and Lincoln (1988), Salvendy (1997), and Helander, Landauer, and Prabhu (1997). Many handbooks are to be found in the automation technology field, such as Shell and Hall (2000). Many of the same topics are covered in handbooks of the traditional engineering disciplines; for example, in mechanical engineering, see Kreith (1997).

Particular industries, or government agencies that support or regulate industries, have often published design guidelines (e.g., see NUREG-0700, U.S. Nuclear Regulatory Commission, 1981, for nuclear plant control room design). In the case of either handbooks or guidelines, the user must be careful to question whether any particular bit of advice, whether implied or stated explicitly, applies in the particular context of the current problem he or she is seeking to solve. Researchers like to generalize their results, and readers often are too ready to accept those generalizations without qualifying them in terms of the particular experiment or design constraints in which they were presented. Sometimes authors preparing design guidelines for one context have simply incorporated into their works the already published guidelines for another design context.

If guidelines are to be adapted from one application context into another, it is mandatory that they be reviewed with the new context in mind. In one case, for a nuclear plant control room alarm, I encountered a government-issued guideline for sound intensity level (in decibels) that seemed rather too loud. Upon investigation I discovered that the specification was lifted verbatim from a recommendation for military aircraft in which the pilots wore earphones and the sound source was on an external panel, so the alarm had to be loud enough to penetrate the sound-barrier ear cups on the earphones. Nuclear plant operators, however, don't wear earphones. The alarm level recommended by the guidelines would almost have deafened these operators.

Standards are a different story. Standards emanate from well-respected organizations such as the International Organization for Standardization (ISO), the American National Standards Institute (ANSI), and the Institute of Electrical and Electronics Engineers (IEEE). Standards committees are formed around particular problem areas by enlisting volunteers from various companies or other organizations representative of that industrial sector. These committees gather in multiple meetings to discuss in technical detail what consensus might be reached on design particulars. When a committee reaches consensus, it publishes a draft standard that is circulated throughout the industry for comment. Based on comments received, the standard is revised, and after several iterations a final standard is published. At that point the standard has real power, because a manufacturer of a product that deviates from the standard may have difficulty getting its products marketed, getting suppliers to carry spare parts, and so forth. Furthermore, tort lawyers can more easily claim so-called design defects if a product's design is shown to have ignored a published industry standard.

Standards making is not always a neat and clean exercise. Stewart (2000) reviewed the status of ISO 9241 (ergonomic requirements for office work with visual display terminals). According to Stewart, users of this standard have had unrealistic expectations that it would solve all their problems. Furthermore, the technology has changed so fast that the committee deliberations have not been able to keep up, and the recommendations that were proffered were often based on the national or industrial interests of the participants rather than the interests of potential users. Nevertheless, in spite of the politics of standards promulgation, product and system designers can obtain much that is useful from standards, along with other information sources such as handbooks, the general scientific literature, and qualified human factors experts.

Chapter 4

HUMAN PERFORMANCE IN RELATION TO AUTOMATION

Automation is usually justified on the basis of improved performance and/or reduced cost. Often the human cost is what the automation proponent has in mind, with the implicit assumption that human involvement can be reduced significantly. This assumption may be naive. For systems involving human safety, the public is typically loathe to do away with the human. Furthermore, the designer may underestimate the costs of human input into design, development, manufacture and testing, installation, maintenance, repair, and management of the automation.

Performance of the human-automation system can be very different from the performance of either element by itself. Evaluation of the automation in strictly technical performance terms is totally insufficient as an evaluation of system performance, partly because many subtle aspects of human decision making cannot easily be evaluated in purely technical terms.

It is just as true that heroic human performance does not necessarily mean superior system performance. A person riding a unicycle may do an impressive job of balancing, but the human-unicycle system is not nearly as fast or as efficient in energy expenditure as, for example, a human-bicycle system.

In this chapter I examine particular aspects of human performance and relate them to that of automation.

Nominal Performance of Humans: Speed and Accuracy

It is clear that human response speed is limited. Imagine the simplest possible sensory-motor scenario: A person holds a hand over a button while watching for a light. When the light comes on, the person pushes the button. Response time is 0.2 seconds minimum, made up partly of nerve conduction time and partly of the visco-inertial lag in activating the hand muscles. Where there is any significant degree of cognitive activity, the response time will be several seconds at least. In a large-scale simulation experiment that tested nuclear power plant operators' ability to observe the displays and make the correct response when alarms suddenly indicated a major loss of coolant, the 95th percentile response time was 100 seconds (Kozinsky, Pack, Sheridan, Vreuls, & Seminara, 1982).

During the 1950s, shortly after Shannon published his groundbreaking paper on information theory (Shannon, 1949), applied psychologists made a great effort to discover at what rate (in bits per second or bits per event) humans could receive information through various sensory-perceptual-memory mechanisms and at what rate they could send information through various motor skill activities.

With respect to sensory-perceptual-memory, it is evident from tachistoscope experiments (visual exposures much shorter than 1 second) that humans can buffer-store complex images long enough to be asked after the exposure what the details were — and then the details are forgotten.

Miller (1956) reviewed the literature on short-term memory for absolute judgments made along simple sensory continua. For example, a participant is presented with N different pitches (or loudnesses, line lengths, light brightnesses, mechanical forces, or degrees of saltiness), each of which (along one of these sensory continua) is assigned a letter (A, B, C, etc.), and the participant tries to remember the assignments. A short time later different items along that continuum would be presented, and the participant would try to name the letter.

Miller found that participants tended to be error free up to $N = 7$ items, but beyond that they made errors. He discovered that the errors increased with more items, such that the transmitted information remained about constant at $\log_2(7)$ bits, slightly more for visual and auditory continua, less for touch, taste, and smell. When multidimensional stimuli were used (e.g., different pitches and loudnesses combined) the information transmitted in the input-output matrix increased in size, but not quite to the level of $\log_2(7 \times 7) = 2 \log_2(7)$ bits.

In the category of sensory-motor skill, Hick (1952) showed that when a participant had to make a different response for each of N different stimuli (e.g., push a different button corresponding to each of N different lights that might come on), the response time was approximately $RT = K_1 + K_2 \log N$.

In Fitts' (1954) experiment with one-axis movements, the participant had to move an object or a cursor a distance A and place it within a tolerance B (in the same direction). He found that response time was approximately $RT = K_1 + K_2 \log(A/B)$.

Robustness, Adaptivity, and Self-Pacing in Continuous Tasks

During the 1960s the U.S. government mounted a major effort to determine the human dynamic equations for target tracking, relating the human response to random-appearing visual input. The primary reason for the interest was to determine the equations (the transfer function) of the human pilot, because in order to do control and stability analysis for aircraft, they had to couple the equation for the pilot with that for the aircraft. The idea was to predict human response as a function of the parameters of the controlled process and the bandwidth of the disturbing signal.

The best-known and most-used model is the *simple crossover model* (McRuer & Jex, 1967). It states that the human adopts whatever transfer function u/e (from observed error displacement e to control device displacement u; see Figure A15 in Appendix A) makes the *forward loop* have a transfer function (from error e to response x of the controlled vehicle or system) be $x/e = (K/j\omega)e^{-j\omega T}$. In other words, the human adjusts his or her own dynamic parameters to make the human-plus-vehicle (forward loop) look like a simple integrator plus a time delay. A simple integrator with high enough gain (see Appendix A, section 5.3) provides good closed-loop control in most cases. The time delay is not desirable, but it is an unavoidable constraint of the human nervous system, T being roughly the same as the discrete reaction time, about 0.2 seconds. This model assumes an error-nulling or *compensatory* display, and it assumes the disturbance does not have significant energy components beyond 1 Hz. It further assumes, according to the equation just given for x/e, that the controlled process is somewhere between a differentiator, for which the human must integrate twice in order to compensate, and a double integrator, for which the human must differentiate once. Most control tasks that a human might be expected to perform are within these restrictions.

As applied to aircraft, this model specifies that for a range of aircraft dynamics, in attempting to keep the aircraft straight and level in spite of random disturbances, the *pilot-plus-aircraft* can be modeled simply as a first-order integrator plus a pure time delay of approximately 0.2 seconds. This is a remarkable finding because it says that the human pilot compensates for whatever aircraft dynamics are imposed to make the combined human-plus-aircraft look like a simple integrator (the human cannot undo the time delay, unfortunately). There are small adjustments to this rule for different dynamics and different bandwidths of the disturbance signal.

A second, *optimal control model* of the human operator (Kleinman, Baron, & Levison, 1970) makes use of the optimal control representation described in Appendix A. It is more robust in some respects but more difficult to use.

All the experimental models just discussed have been repeated and refined over recent years but stand as the classical representations of human response speed and accuracy as a function of task complexity.

It should be emphasized that there are limits to what a person can control. If the disturbance has bandwidth greater than 1 Hz, the human cannot keep up and produces mostly just noise not coherent with the disturbance. If the dynamics are undamped with a denominator greater than second order (more than a second integrator) or a numerator more than first order (a differentiator), the human cannot maintain stability. Most real systems fit this range of parameters, and both the McRuer and Jex (1967) crossover model and the Kleinman et al. (1970) optimal control model have proven to be very useful generalizations — all the more impressive because they are among the few quantitative models that exist for human-plus-machine.

If each successive stimulus does not demand instantaneous response as soon as it is seen, humans wisely do *self-pacing*. For easy, less complex tasks (fewer variables or dimensions, fewer discriminations per variable) they speed up, and for more complex tasks they slow down. When driving cars, people tend to speed up on the straight roads and slow down on the curves, thereby reducing maximum information rate or bandwidth.

For cognitive tasks, people plan ahead and thereby reduce information overload and the mental stress that goes with it. Pilots have the phrase, "keeping ahead of the airplane." At the same time, people do not pace their work and their lives to monotonous sameness. We seem to like some variability.

Biases of Human Decision Makers

There are large individual differences between decision makers. These differences can be differentiated in several dimensions (Pew and Mavor, 1998):

1. *Degree of decision-making experience*. This includes sophistication with respect to concepts of probability, and how probability combines with level of consequences to form a construct of risk or safety.
2. *Acceptance of risk*. Most decision makers are risk-averse, whereas some are risk-neutral or even risk-prone.
3. *Relative weighting on cost to self* as compared with cost to some adversary or other party.
4. *Tendency to decide on impulse* as contrasted with deliberation.

None of the foregoing can be called *irrationality;* they are just differences in decision-making style.

Earlier studies of decision making compared empirical behavior with the norms established in classical decision theory (Einhorn & Hogarth, 1978; Edwards, 1954; Slovic, Fischhoff, & Lichtenstein, 1977). There was consensus that people are fairly good at estimating means and variances, and also proportions — unless probability is close to 0 or to 1, and then they are not. People tend to regard very large numbers, such as 10^5, 10^6, 10^7, and 10^8, as the same, even though the numbers may be orders of magnitude different from one another. This may also be said for very small numbers, such as 10^{-5}, 10^{-6}, 10^{-7}, and 10^{-8}. This surely results in part from a lack of personal experience with such numbers. It may also be attributed to the empirical fact that sensation is logarithmically related to the physical stimulus. People are much better at making ratio comparisons between large numbers and small numbers when the ratio is not greater than 1000. Winkler and Murphy (1973) showed that weather forecasters are one of few groups that are good at quantitative estimation.

Accumulated recent research has made it clear that humans systematically deviate from rational norms such as the Bayesian probability updating model (see Appendix A). Deviations from the Bayesian model include the following empirical facts:

a. Decision makers do not give as much weight to outcomes as Bayes' rule would indicate (Edwards, 1968). Probabilities of alternative hypotheses or propositions tend to be estimated much more conservatively than Bayes' theorem of probability updating would predict.

b. They tend to neglect base rates (Edwards, 1968; Tversky & Kahneman, 1980). This is the common tendency to overweight recent evidence to the neglect of previous evidence. Of course, for a rapidly changing situation (a nonstationary statistical process) this may be rational.

c. They tend to ignore the reliability of the evidence (Tversky & Kahneman, 1974).

d. They are not able to treat numbers properly as a function of whether events are mutually independent or dependent. They tend to overestimate the probability of interdependent events and underestimate the probability of independent events (Bar Hillel, 1973).

e. They tend to seek out confirming evidence and disregard disconfirming evidence (Einhorn & Hogarth, 1978).

f. They are overconfident in their predictions (Fischhoff, Slovic, & Lichtenstein, 1977).

g. They tend to infer illusory causal relations (Tversky & Kahneman, 1973).

h. They tend to recall having had greater confidence in an outcome's occurrence or nonoccurrence than they actually had before the fact (Fischhoff, 1975). This is called *hindsight bias*: "I knew it would happen."

Tversky and Kahneman (1974) showed that the foregoing discrepancies can be attributed to three heuristics:

1. *Representativeness.* The probability of an event's belonging to a category B is judged by considering how representative A is of B. Because long series of heads or tails when flipping coins is considered unrepresentative, people are likely to predict the other event on the next trial. The illusion of validity occurs when people treat highly correlated events as though they are independent, thereby adding to the weight of one hypothesis.
2. *Availability.* An event is considered more likely if it is easy to remember; for example, an airplane crash. Evans (1989) argued that people's consideration of available but irrelevant information is a major cause of bias.
3. *Anchoring and adjustment.* This is the idea that people update their degree of belief by making small adjustments (based on the new evidence) relative to the last degree of belief (Hogarth & Einhorn, 1992).

Representation or *framing* effects, sometimes also called *prospect theory* (Tversky & Kahneman, 1992), deals with how situations are presented to the decision maker, and the tendency to see situations in terms of hopeful possibilities. Judgments can be different depending on how the question of proposition is framed, even though the probabilities and consequences remain the same. For example, in medical interventions, the results are different depending on whether the outcomes are posed as gains or losses (Tversky & Kahneman, 1981). People will overestimate the chance of a highly desirable outcome and underestimate the chance of an undesirable one (Weber, 1994). A food's being "90% lean" is more acceptable than if it is "10% fat." Fifty percent survival is more acceptable than 50% mortality. In other words, a glass half full is more believable than a glass half empty.

Outcome evaluations are often relative rather than absolute. For example, deciding on the basis of *regret* refers to seeing the outcome relative to what might have been. Purchasing a failing stock is seen as worse than not selling the same stock when already owned (Kahneman & Tversky, 1982). Deciding on the basis of sunk cost refers to persisting in a failing action to which one has made an initial commitment. For example, people who paid full price for a ticket are more likely to attend a theater performance than if they unexpectedly receive the tickets at discount price (Arkes & Blumer, 1985).

Situation Awareness

Situation awareness (SA) is a popular term in aviation and other communities concerned with human safety in dynamic situations. It has been variously defined. There is moderate agreement that it refers to an operator's state of knowledge of the relevant aspects of a dynamic environment with which a person is interacting. Endsley (1995) defined it as "the perception of the elements of the environment within a volume of time and space, the comprehension of their meaning, and the projection of their status in the near future" (p. 65).

SA can be critical when the human operator is confronted with a complex and changing situation. For example, Endsley and Strauch (1997) claimed that in the 1995 B-757 crash near Cali, Columbia (discussed in Chapter 2), the Flight Management System confused the crew, led to loss of situation awareness, and caused the accident.

There is little consensus on how to treat SA theoretically. It is obvious that SA is affected by many other cognitive factors, such as attention and distraction (one cannot be aware of something if one has not attended to it or is distracted from attending to it). SA is also affected by motivation (one attends to and remembers what one cares about) and by workload and stress (Endsley & Kaber, 1999). One is limited in one's capacity to attend to or remember if one is otherwise occupied (Neisser, 1976). Others have suggested that knowledge, schemata, or mental models (how information is represented in the head) lead to anticipation of certain events and thus direct exploration and attention (activity to acquire new information), which in turn affects the sampling of the available information in the environment and determines what new information is available for modeling, which modifies the schema and mental models in the head, and so on in a cycle. Figure 4.1 depicts my representation of this cycle.

In an effort to dispel what they called the "knowledge versus process dilemma," Smith and Hancock (1995) considered SA to be "adaptive, externally directed consciousness" (p. 137). Flach (1995) warned of the

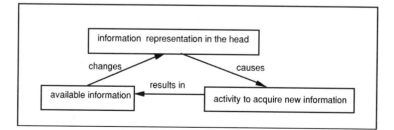

Figure 4.1 Neisser's perceptual cycle.

pitfalls of treating SA as both cause and effect (though one can assert that behavior, being part of a closed loop, is unavoidably both).

Measurement of SA is most easily done in a simulator by randomly and unexpectedly (to the experimental participant) freezing the scenario and asking questions about the state of key variables (or, for example, by having the participant place tics on multiple numerical scales, maps, etc.). Endsley (1995) called this the *situation awareness global assessment technique* (SAGAT). SA can also be measured by ongoing subjective ratings or debriefing after a real or simulated event, but the problem is that the participant may have forgotten or have had his or her recall corrupted by interceding events.

An interesting question is how SA can be enhanced, either by self-imposed mental procedures or by external means. Normally, as behavior becomes "automatic," people may pay attention to salient environmental elements when necessary, but they may fail to remember the details because they were essentially subconscious. For example, in driving to work in the morning, one would almost surely fail to notice if a speed limit sign had a different posting. People don't attend to things they don't expect. For this reason, at least partly, pilots are taught to go through pretakeoff and prelanding checklists and force themselves to take notice of the readings of certain instruments. In Japan, train drivers are taught to point at each signal to ensure that they give it proper attention.

External means to enhance situation awareness may involve the use of warnings (if the computer detects some variable that is troublesome or moving in that direction) and computer-generated reminders based on a schedule or phase of a mission. It is important to use situation awareness concepts and measures for evaluating proposed displays and decision aids.

Trust

Trust is now considered an important factor in human-automation performance. The term can have different meanings to different people (Sheridan, 1988). It can be both an effect and a cause. As an effect, it can mean judged reliability of the system, in the usual sense of repeated, consistent functioning. It can mean perceived robustness, demonstrated or promised ability to perform under a variety of circumstances. It can mean familiarity — that is, the system employs procedures, terms, and cultural norms that are familiar, friendly, and natural to the trusting person. It can mean understandability in the sense that the human supervisor or observer can form a mental model and predict future system behavior. It can refer to the system's explication of its own intention, how it explicitly displays or conveys that it will act in a particular way, as opposed to requiring that its future actions be predicted from a model. Finally, it can refer to usefulness, or utility of the system to the trusting person in the formal theoretical sense. It can simply mean dependence of the trusting person, whether desired or not.

Trust can also be a cause. The human's use of the automation depends on his or her trust of it. If it is not trusted, it will be avoided. Human operator trust in automation is now a major topic of interest (Muir & Moray, 1996; Riley, 1994; Wickens, 1994) because it significantly affects whether and how automation is used.

Lee and Moray (1992) studied the emergence and decay of trust of automation in supervisory control of a simulated pasteurizing plant, with optional use of automation (or reversion to manual control) and randomly introduced automation failures. They suggested a quantitative model of dynamic changes in operators' trust and their decisions to use the automation, as a function of past level of trust, failure probability, and capability in manual control. Shifts between automatic and manual control modes can be predicted by the ratio between trust in the machine and self-confidence (Lee & Moray, 1992, 1994; Muir, 1988).

Engendering trust is often a desirable feature of a system, something the designer strives for — but not always. Too much trust (usually naive trust) can be just as bad as too little. Parasuraman and Riley (1997) made a compelling case that system designers should be concerned about misuse, disuse, and abuse of automation, based on distrust and overtrust as well as workload and other factors.

Mental Workload

Mental workload has been a popular topic in human-machine systems research for the past two decades. Much of this research has focused on pilots, partly to resolve disputes about whether the commercial aircraft flight deck should have a crew of two or three, partly to decide on the limits of what can be assigned to fighter and helicopter pilots flying nap-of-the-earth combat missions. It has also been of concern in nuclear power plant control rooms and other settings where overload and concomitant deterioration of operator performance pose a serious threat.

Moray (1979) provided a number of views of mental workload, including both its definition and its measurement. Hart and Sheridan (1984) reviewed the definition and measurement of workload as regards automation. They pointed out that whatever *mental* workload is,

- Mental workload is not *physical workload,* the mechanical energy expended (e.g., in calories) by the operator in doing the assigned task, measured by the heat given off or by the conversion of respired oxygen to carbon dioxide. It is true that positron emission technology scans reveal more biochemical activity in active parts of the brain during cognitive activity, but the energy levels of brain tissue are insignificant compared with those for muscle tissue.

- Mental workload is not an objective measure of what task is to be done or how complex the task is. *Task load* is a property of the task, independent of the operator or the operator's behavior. Examples of task load are the number or variety of stimuli (displays, messages) or actions required, rate of presentation of different displays or required pace of actions, time urgency before a deadline, improbability of the information presented or the actions required, and degree of overlap of multiple-task memory demands. An experienced operator is likely to be able to perform a task well with little mental workload, whereas an inexperienced operator may perform poorly with great mental load. A robot may be said to perform a task with a given task load with zero mental workload.

- Mental workload is not human operator performance. *Operator performance,* which translates to system performance and safety, is surely the ultimate concern. It is not mental workload per se that is of ultimate concern. It is whether the pilot or other operator can perform his or her job with any margin of attention for unforeseen circumstances, stress, or complexities, which may be infrequent but certainly do arise from time to time. Until mental workload reaches a high level, however, operator performance (or system performance) may show little decrement. Then, with a slight further increase in workload, performance may decrease precipitously (Figure 4.2). The major assumption is that indices of mental workload can provide a better predictor of performance breakdown, predicting the breakdown well before it occurs, better than direct measures of performance itself.

Measures of Mental Workload

The following are the three major measures (and in that sense also definitions) of mental workload:

- *Physiological indices* used for mental workload assessment are numerous. Some of these are heart rate, heart-rate variability, changes in the electrical

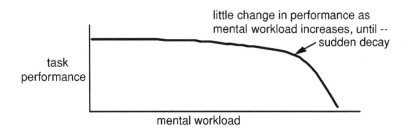

Figure 4.2 *Performance decay as a function of mental workload.*

resistance of the skin caused by incipient sweating, pupil diameter, power spectrum changes of the voice (especially increases in mean frequency), and changes in breathing pattern. Great variability has been found in these measures; no single measure has been accepted as a standard.

- *Secondary task performance* measurement requires that the experimental participant do a secondary task at the same time he or she is performing the primary task. Examples of secondary tasks are backwards counting by threes, generation of random numbers, and simple target tracking. The argument goes that the better performance is on the secondary task, the less the workload is of the primary task. The problem has been that airplane pilots and operators performing actual critical tasks sometimes refuse to cooperate, although in simulators (where they know there are no real dangers) they may be cooperative.

- *Subjective rating* is by definition a nonobjective measure, but, interestingly, it is the standard against which all objective measures are compared. Subjective mental workload rating can be done according to a single-dimensional or a multidimensional scale.

Sheridan and Simpson (1979), after systematically observing pilot behavior on a number of airline flights, proposed that the Cooper-Harper 10-level category scale that had previously been used for test-pilot rating of aircraft handling qualities be adapted for subjective scaling of mental workload. An experimentally validated version of such a scale was developed by Wierwille and Casali (1983); see Table 4.1.

Sheridan and Simpson also suggested a multidimensional scale using separate dimensions of time-pressure ("busyness"), mental effort (problem complexity), and emotional stress. One such scale, developed by Reid, Shingledecker, and Eggemeier (1981), is shown in Table 4.2. A similar scale, developed for the U.S. Air Force by O'Donnell and Eggemeier (1986), the Subjective Workload Assessment Technique (SWAT), is now commonly used. Another popular scale, developed for NASA, the Task Load Index (TLX), asks the respondent for ratings along six axes: mental demand, physical demand, temporal demand, performance, effort, and frustration level (with questions to help define the meaning of each category).

Underload and Workload Transients

Common sense dictates that if human operators are given too little to do, they will become inattentive and their monitoring performance will decline. The classical literature on vigilance generally supports this claim (Buckner & McGrath, 1963). If there is too much workload, performance also declines. Many researchers have been concerned that there must be an

TABLE 4.1
Subjective Scale of Mental Workload.

Operator Demand Level	Rating
Operator mental effort is minimal, and desired performance is easily obtainable	1
Operator mental effort is low, and desired performance is attainable	2
Acceptable operator mental effort is required to attain adequate system performance	3
Moderately high operator mental effort is required to attain adequate system performance	4
High operator mental effort is required to attain adequate system performance	5
Maximum operator mental effort is required to attain adequate system performance	6
Maximum operator mental effort is required to bring errors to moderate level	7
Maximum operator mental effort is required to avoid large or numerous errors	8
Intense operator mental effort is required to accomplish task, but frequent or numerous errors persist	9
Instructed task cannot be accomplished reliably	10

Adapted from Wierwille and Casali (1983) with permission.

optimum somewhere between the extremes of task load. However, finding where such an optimum exists for any particular task has been an elusive goal.

Aircraft piloting, monitoring of plant operation, and military command and control can involve long periods of boredom punctuated by sudden bursts of activity, without much demand for activity at the much-preferred intermediate levels. To some extent this is true even if no automation is present (e.g., long periods of manual piloting to maintain course), but automation only exacerbates the problem. It is not that the operator is simply bored doing a manual task. He or she may not be involved at all for long periods and may become "decoupled" even from feeling responsible for control, knowing a computer is doing the job. However, when problems arise that the automation cannot handle, or the automation breaks down, the demands on the operator may suddenly become severe.

Such workload transients — coming into the game cold from the bench, so to speak — pose serious risks of human error and misjudgment. Experienced operators will react instinctively and often be correct, or if the situation is novel, they may try to buy time to collect their wits. Operators have been known to accommodate to information overload by bypassing quantitative input channels (e.g., instruments) and going instead to ad hoc qualitative channels (e.g., voice messages), which may not be reliable.

TABLE 4.2
Three-Dimensional Mental Workload Scale

Time Load	Rating
Often have spare time. Interruptions or overlap among activities occur infrequently or not at all.	T1
Occasionally have spare time. Interruptions or overlap among activities occur frequently.	T2
Almost never have spare time. Interruptions of overlap among activities are very frequent or occur all the time.	T3

Mental Effort Load	Rating
Very little conscious effort or concentration required. Activity is almost automatic, requiring little or no attention.	E1
Moderate conscious mental effort or concentration required. Complexity of activity is moderately high because of uncertainty, unpredictability, or unfamiliarity. Considerable attention required.	E2
Extensive mental effort and concentration are necessary. Very complex activity requiring total attention.	E3

Stress Load	Rating
Little confusion, risk, frustration, or anxiety exists and can be easily accommodated.	S1
Moderate stress because of confusion, frustration, or anxiety noticeably adds to workload. Significant compensation is required to maintain adequate performance.	S2
High to very intense stress caused by confusion, frustration, or anxiety. High to extreme determination and self-control required.	S3

Adapted from Reid et al. (1981) with permission.

Automation has been motivated by a desire to reduce workload, in some cases by people oblivious to the fact that the requirement for human monitoring of the automation itself creates workload (hence the pilot's term, "killing us with kindness").

Wierwille, Casali, Connor, and Rahimi (1985) reported a series of evaluations of the sensitivity and intrusion of mental workload estimation techniques. In simulated aircraft piloting they imposed psychomotor tasks (wind disturbances to be nulled), perceptual tasks (detection of instrument failures), cognitive tasks (wind triangle navigation problems), and communication tasks (execution of commands, response to queries). They measured subjective workload ratings, several attributes of primary task performance, several aspects of secondary task performance, and various physiological indices.

Wierwille et al. found that subjective ratings were highly sensitive to changes in imposed workload. Primary task performance measures (e.g., aircraft response) were not sensitive, except for those measures that corresponded to the operator's control movements. Time from presentation to response was sensitive to the perception and cognitive tasks, even more so than the subjective ratings (complex demands simply took longer to fulfill). Conversely, communication response times were shorter for increased task load. Most secondary task measures fared poorly (e.g., mental arithmetic, memory scanning, regularity of tapping), except for the standard deviation of time estimation. Secondary task measures cannot be used simultaneously with subjective ratings, as the latter are affected by both primary and secondary tasks. Physiological indices (pulse rate, respiration rate, pupil diameter, eye blinks, and eye fixations) were mostly insensitive to imposed task load.

Young and Stanton (1997) explored the problem of how driver workload is affected by the increased automation being introduced into the automobile.

Stress and Fatigue

Closely associated with mental workload are stress and fatigue. Both phenomena are usually studied in physiological terms, as measured by subjective symptoms and body chemistry. Fatigue itself is clearly associated with loss of sleep, circadian rhythm (usually measured by body temperature and ease of falling asleep), and muscular effort expenditure. Neither stress nor fatigue is particularly well defined, either as a cause or as an effect. Hancock and Desmond (2001) reviewed stress and fatigue along with workload in aviation and highway driving.

Human Error

What is an error? Are errors caused? If so, what are the causes? How many causes are there — an infinity? Do errors occur at random? Can they be predicted? Would it be desirable to eliminate all errors, or are errors part of creativity? What is the relationship of an error to an accident? To a fault or a failure? To motivation, to seriousness of consequences, to crime, to mental disease, to sin? Are humans who interact with automation systems more or less prone to error than otherwise? Does or can the automation compensate for human error, or does the human-automation interaction exacerbate the error?

On one hand, interest in human error at the political, legal, and everyday levels has always been strong. Scientific literature on human error, on the other hand, is woefully scant in relation to its importance. Behavioral science has focused mostly on *correct* behavior and not much on *incorrect* behavior. Nuclear power and commercial aviation regulators have been particularly interested in human error in recent years because of public interest in accidents. Ruffell-Smith (1979) studied pilot errors and fault detection under heavy cognitive load and found that crews made approximately one error every 5 minutes. The Institute of Medicine report *To Err Is Human* (Kohn, Corrigan, & Donaldson, 1999) has aroused the whole medical community to the seriousness of human error in diagnosis and treatment in hospitals. There are a few excellent books on human error generally, including Reason (1990, 1997), Senders and Moray (1991), and Perrow (1984).

Problems with the Definition of Error

Human error is sometimes defined as an action that fails to meet some implicit or explicit standard of the actor or of another person. Such a definition, unfortunately, begs the question of what is the standard. The point here is that "error/no error" is the simplest possible (binary) categorization of complex human behavior and that such a one-bit decision depends on a standard that may be considered arbitrary. Any behavior can be relegated to error or not by modification of the standard. It seems that a lack of consensus on a definition is the greatest source of variability in both modeling and empirical measurement of human error (Sheridan, 1983).

A comparison with other forms of human endeavor is interesting. Ordinary human discourse tolerates infinite variation and shades of inference about human behavior. Psychiatry seeks at least qualitative categories of behavior. Psychometrics requires continuous quantitative scales of behavior. In contrast, human error analysis typically reduces behavior to a single binary discrimination. Varying degrees of human error are usually disallowed by reliability analysts.

Operator error count (as used for human-machine system reliability analysis) may be more a function of the measurement criterion of the analyst than of the behavior of the operator or the system. For example, it is common in nuclear power plants that operators perform procedural steps within certain groups in an arbitrary order because they know that the order of steps within these groups "doesn't matter." Some steps may be delayed, omitted altogether, or performed to a different criterion because that's the way a particular operator was trained or understands his or her task. In human reliability studies, however, because experimental protocols used criteria that were different from those familiar to the participants, these have been scored as errors.

It is common for continuous variables to be counted as errors if they exceed some fixed value. A human-controlled continuous variable may make a single slow excursion just across the limit and back and be counted as one error, it may cross just over and back several times and be counted as several errors, or it may hover just short of the criterion and not be counted at all. To the experienced operator, these three situations may be considered equally bad, but again, the criterion determines the error count. An operator may allow a variable to trigger an alarm knowing full well that he or she can easily recover, but nevertheless the operator is charged with an error.

Some occasionally speak of "good errors." The control engineer would assert that there can be no feedback control without an error signal — a measured deviation, however small, from a desired reference. The learning psychologist would assert that error is part of learning and skill development. The artist would claim that error is essential to creativity. Darwin claimed that error (what has come to be called *requisite variety*) is an integral part of evolutionary improvement of living creatures.

Error Taxonomies

Many taxonomies or classification schemes exist for human error. Common distinctions are the following:

- errors of *omission* versus errors of *commission;*
- errors in *sensing, memory, decision*, and *response;*
- errors in deciding what one intends (an incorrect intention is called a *mistake)* versus errors in implementing those intentions (an execution not in accord with one's intention is called a *slip;* see Norman, 1981); and
- *forced* errors, in which task demands exceed physical capabilities, versus *random* errors, which can be slips or mistakes.

Reason (1990) proposed the classification scheme pictured in Figure 4.3. Unsafe acts are unintended or intended. The unintended acts can be slips or lapses, whereas the intended acts can be mistakes or purposeful violations. Slips, lapses, and mistakes are errors.

Causes of Human Error

It is easy and common to blame operators for accidents, but investigation often suggests that the operator "erred" because the system or the procedures were poorly designed. That errors have causes seems obvious. Still, investigations of errors or accidents seldom come up with neat explanations of causality (unless they expediently truncate their investigation with simplistic explanations such as "driver drunk" or "inattention"). Consider the typist who has hit an incorrect key. What is the cause? Most behavioral scientists assert there is no one absolute cause but, rather, something closer to a causal chain leading to the error. In accident situations there is likely to be a causal chain leading from error to accident. The

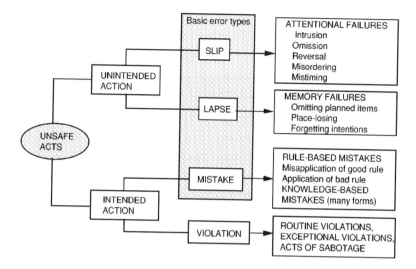

Figure 4.3 Reason's (1990) taxonomy of human errors. (Reprinted with permission.)

hope, then, is to be able to anticipate what Reason (1990, 1997) called "latent" errors and what Perrow (1984) called "normal" accidents, to intervene in the chain, preferably before the error but certainly before the accident. It has been said that whereas causes somehow may make errors happen, "reasons" are how people rationalize later. The following are some of the popular causes — or reasons.

Lack of feedback. The idea is that, as in machines with closed-loop or feedback control, human error is simply an epiphenomenon of inadequate feedback (measurement or sensitivity), so that the system state wanders into an unacceptable region of the state space. Lack of feedback is probably the most important cause of error. Feedback (or lack thereof) is determined not only by what is made available by the automation or otherwise exists in the physical environment, it is also determined by whether and how the human accesses it. Thus loss of attention or situation awareness, or confusion over which mode the automation has been set to, also determines what feedback reaches the human (Sarter et al., 1997).

Capture. In complex tasks with a variety of procedures, a frequent cause of a slip error is *capture*. Assume that the operator is well practiced in some task sequence *ABCD* but occasionally intends to do *EFBG*. Both sequences require the common element *B*. The operator does *EFBCD* and then discovers the error (Figure 4.4). Somehow the operator's conditioning to do *C* after *B* was overwhelming. The correct but abnormal *B–G* course was not sufficiently compelling at the critical time.

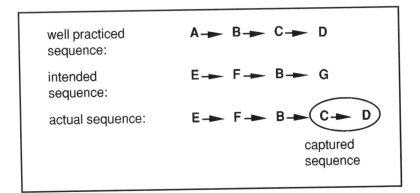

well practiced
sequence: A→ B→ C→ D

intended
sequence: E→ F→ B→ G

actual sequence: E→ F→ B→(C→ D)

                              captured
                              sequence

Figure 4.4 Capture error.

Invalid mental models. One emerging theory of error causation appeals to the idea of a mental model, an input-output cognitive representation of the controlled process. For given process input, the mental model can be used by the human brain to observe, or infer by cognitive simulation, variables that are not convenient to measure directly with the senses.

Hypothesis verification and the law of small numbers. Much laboratory research on individual decision making has shown that participants work to verify hypotheses they hold, searching for and retaining confirming evidence and ignoring or forgetting contradictory evidence and knowledge of what has not failed (Rouse & Hunt, 1984).

Stress and perceptual narrowing. A commonly observed concomitant of stress is *perceptual narrowing*, also called *tunnel vision* or *cognitive lockup*, meaning the tendency to limit one's physical or mental attention and action to what is most immediate and familiar, being unable or unwilling to avail oneself of a broader set of options.

Risk (error) homeostasis. This is the notion that people inherently tend toward some level of risk (for whatever genetic, psychological, or socio-logical reason) and therefore actively subject themselves to disturbances to their (otherwise) condition of stability.

State of the nervous system. Blood alcohol level, drowsiness, and emotional stress affect the nervous system and are also correlated with error.

"Latent" property of human error. Reason (1994) emphasized the latent property of human error as follows:

> Unsafe acts are like mosquitoes. You can try to swat them one at a time, but there will always be others to take their place. The only effective remedy is to drain the swamps in which they breed. In the case of errors and violations,

the "swamps" are equipment designs which promote operator error, bad communications, high workloads, budgetary and commercial pressures, procedures that necessitate their violation to get the job done, inadequate organization, missing barriers and safeguards . . . . the list is potentially very long, but all these factors are, in theory, detectable and correctable before a mishap occurs. (p. xiv)

At present there is no acceptable unified theory of causation of human error.

Remedies for Human Error

Given some understanding of the error situation, the usual wisdom proffered by the human factors profession for keeping so-called bad errors in check is (in order of efficacy) as follows.

Design to prevent error. Provide immediate and clear feedback from an inner loop early in the consequence chain. Provide special computer aids and integrative displays showing which parts of the system are in what state of health. Give attention to cultural stereotypes of the target population — for instance, in Europe the expectation is that flipping a wall switch down turns a light on, so when designing for Europeans, don't use the American stereotype. Use redundancy in the information, and sometimes have two or more actions in parallel (though this does not always work). Design the system to forgive, and to be "fail safe" or at least "fail soft" (i.e., with minor cost).

Train. Teach operators the mental models appropriate for their tasks, and make sure their mental models are not incorrect. Get operators to admit to and think about error possibilities and error-causative factors, because although people tend to catch errors of action, they tend not to catch errors of cognition. Train operators to cope with emergencies they haven't seen before, using simulators where available. Use skill maintenance for critical behaviors that need to be exercised only rarely.

Restrict exposure to opportunities for error. Make the fire alarm or the airplane exit door have two or more steps so that actuation is not inadvertent. Have key locks for certain critical controls that are seldom required so that inadvertent actuations do not occur. However, be conscious that this limits the operator's opportunity in crisis.

Alarm or warn. Keep in mind that too many alarms or too many warnings, whether on the control panels or in printed instructions, tend to overload or distract the observers and condition them to ignore them. Tort lawyers would have everyone believe that warnings are the most essential means to ensure safety. They may be the best way to guard against lawsuits, but they are probably the least effective means to achieve safety from a human factors viewpoint.

Consider what behavior is acceptable, what errors are likely, and what to do about them. It is best to strike a balance with tolerance of variability and not expect people to be error-free zombie automatons. If automation is indicated, try to keep the operator knowledgeable about what the automation is doing, provide opportunity for takeover from the automation if it fails, and engender some responsibility for doing this. Don't be too quick to blame the operators closest to the apparent error occurrence. Tilt toward blaming the system, and be willing to look for the error latencies.

Human Error and Automation

By automating, engineers hope to avoid human error. To some extent that works, so long as the automation does not fail and is sufficiently robust to handle all circumstances. However, human operators are necessarily put in charge of setting up the equipment, programming the computers, monitoring their operation, detecting and diagnosing failure, and, if failure occurs, setting things right. In all of these stages human errors occur.

Examples of human error in various automation contexts were given in Chapter 2. It should be noted that the same phenomena that cause human error without automation also cause error when automation is present. The automation, insofar as it allows for human interaction with it, also allows for human error to affect it. In fact, with automation the human operator is sometimes less able to recognize that he or she has made an error, because the automation accepts the human input and then is more opaque than if that input were one of direct and continuous manual control.

The following summarizes the events leading to the crash of a $3.7 million Air Force Predator unoccupied air vehicle (UAV) in Nevada in 2000. During a routine flight the pilot, who was very experienced with this UAV, was trying to adjust signal strength settings to improve his video and telemetry quality. When he was finished, though, he didn't return to the normal command menu. Except for a different border color, the extended menu button interface was visually identical to the normal menu button interface. However, the keyboard's function keys did different things in each menu. Because the pilot was in the wrong menu, when he pressed a key intending to bring up another submenu, he inadvertently depressed the wrong button. That led the control system to dump essential preprogrammed data so that the drone lost its data link connection with the ground control station from which the pilot was operating. The UAV, instead of transitioning to a stable lost-link profile, stalled and crashed 3 minutes later. Investigators noted that the pilot had gotten into the habit of using the extended menus to fine-tune his systems, a procedure that was contrary to their training.

This is a classic case of what has come to be called a *mode error.* There is consensus among human-automation designers that we should be designing our control interfaces, especially with respect to computer commands and displays, so that they are more transparent to the actual working system. The operator should easily "see through" the displays to what is going on and anticipate what will occur shortly. Operators want to know what the automation is doing and what will it do next. They do not like to be surprised.

All human errors associated with automation are not so complex. Much simpler forms of automation can be ill-designed and/or misunderstood by humans who come up against them. In one rather amusing case, an Air Force hangar containing expensive combat-ready aircraft was rigged with light- and heat-sensitive gear designed to detect fires and then flood the hangar with foam if triggered. The system was armed each night after everyone left for the evening. However, one janitor thought the airplanes looked pretty impressive in the hangar and decided to return at night and take some photos. The system triggered as designed and blanketed the hangar with foam. The janitor doesn't work there anymore.

Human-Machine System Reliability Analysis

Humans and machines differ. The Fitts list reviewed earlier, plus common sense, tell us that humans and machines are quite different in their capabilities and the way they fail. Humans are erratic, and they err in unexpected ways. Machines are dependable, which means they are dependably stupid when asked to perform what is beyond their capability. Humans are conscious of their own capabilities and can often recognize unexpected situations and ask for help when needed. Machines are seldom equipped to stop a minor error from propagating. Humans can recover from their own errors, whereas machines seldom can.

Various methods have been offered for analyzing erroneous human behavior. For example, Cacciabue (1997) presented a method that includes a taxonomy of errors, a set of data and correlations from the working environment (he has been concerned mostly with nuclear plants and aviation), and a procedure.

A principal culprit in the management of complex automation is that subsystems come from many different suppliers, and often each supplier provides special alarms and controls as though that subsystem can operate independently of the others. Also, the supplier naively assumes that the human operator has sufficient attention to devote to those alarms and controls. The system designer may defer to the supplier as to what is best for that subsystem. As a result, in an emergency, there can be an excessive number of alarms and special control requirements that cannot be heeded.

Statistical analysis is now the favored diagnostic tool in aviation, nuclear power, medicine, and other fields to reveal cause-effect relationships resulting in safety breakdown. However, according to Maurino (2000), statistical analysis does not reveal the underlying processes that must be understood to make system changes and which are heavily influenced by operational context and organizational culture. Maurino emphasized the need to stay away from the "blame and train" cycle, which has been so common in the past. He emphasized (as do most thoughtful writers on the topic) that safety is a social construct, different within every culture, and that acceptable risk is a subjective judgment, not inherent in the human-machine system. Thus it is naive to export a safety solution that worked in one culture and expect it to work in the other. The organizational culture will determine what is acceptable (Helmreich & Merritt, 1998). A safe culture, Maurino claimed, requires that decision makers place a strong emphasis on identifying hazards and controlling risks and at the same time communicate realistic views and foster a climate of receptivity to criticism or other feedback from lower levels of the group.

Conventional error analysis backtracks a sequence of events until an event is discovered that may be considered "bad," and that is taken as the cause. This often fails to consider the motivations and other circumstances that led to the event or prior errors (or nonerroneous symptoms) that did not seem in themselves to be sufficiently bad. Unfortunately, most accident investigation agencies pay only lip service to human factors, either from ignorance or for convenience.

Given these caveats, I will review four methods of combining human errors and machine failures to determine system reliability.

Combinatorial Analysis of Failures of Particular Events

By this technique one estimates the probability of a particular system function. It uses conventional combinatorial mathematics for independent elements — namely, the probability of success of two (or more) elements (human or machine) in series is the product of their individual probabilities of success. For three elements, $P(A,B,C) = P(A)\,P(B)\,P(C)$. However, if they are not independent, $P(A,B,C) = P(A)\,P(B|A)\,P(C|A,B)$. In the latter case the contingent probabilities are more difficult to assess, especially when multiple contingencies are involved. Note that if the data available are probabilities of failure, they must be converted — that is, $P(\text{success}) = 1 - P(\text{failure})$.

The analyst makes use of two graphical tools, *event trees* and *fault trees*. Figure 4.5 gives an example of an event tree for seeing a colleague at a meeting. Each of the blocks in the vertical series contains a probability, which in this example is most likely independent of the others, so the success probabilities can be multiplied. Note that when the tree diverges, the probabilities at the same horizontal level are mutually exclusive so that their sum must be one.

Figure 4.6 gives an example of a fault tree. Again, each block is represented by a probability, in this case a probability of failure. The probabilities here are determined by whether the logical relation of inputs is AND or OR. An AND for any hypothetical independent inputs A and B means that the probabilities are multiplied to determine the successor probability; that is, if both inputs must fail for the output to fail, P(output failure) is $P(A$ fail)$P(B$ fail). With an OR, either input (A fail) or (B fail) by itself, or both failing together, will result in output failure, and the only other possibility is that neither fails. So P(output failure) is $1 - P(\text{neither}) = 1 - [1 - P(A)][1 - P(B)]$. Such combinatorial analysis eventually determines the probability of the "final event," the one of primary concern. Final events are such things as failure of a whole plant to generate power, significant release of radiation, or crash of a vehicle. The nuclear power and defense industries have made major use of such techniques (e.g., see Meister, 1964; Swain & Guttman, 1983). Note that event trees diverge with

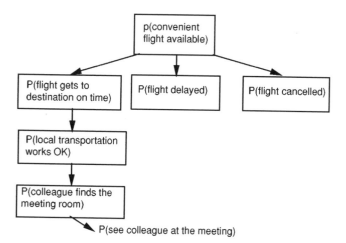

Figure 4.5 Event tree for P(seeing colleague at a meeting).

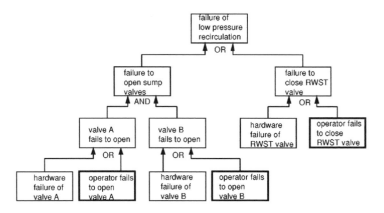

Figure 4.6 Example of an actual fault tree used by nuclear safety analysts. Human elements are blocks with heavier lines.

time and causality, whereas fault trees converge. In addition, for both event trees and fault trees, both human and machine elements can be put into the boxes.

Sensitivity Analysis

Sensitivity analysis refers to studying how changes of some input affect some output of interest. Two kinds of sensitivity analysis are mentioned here in the human-machine reliability context. In one a complex mathematical model exists on a computer and the analyst runs the model over and over while systematically examining the effects of changes in one or another variable. For example, Hall, Samanta, and Swoboda (1981) analyzed how changes in human error probabilities (HEPs) might affect system unavailability, core melt probability, and radiation release probability. They started from nominal probabilities for salient events and varied HEPs by a variety of factors, both smaller and larger than an accepted standard value (such as might result from better training or, on the other side, worse motivation). They showed that in some cases HEPs could become very large but made little difference in safety because machine elements took over, whereas in other cases a small change in HEP was catastrophic.

A second type of sensitivity analysis is largely based on expert subjective opinion. In this case *performance-shaping factors* are devised in order to specify how HEPs change as a function of all the various conditions of work and rest cycles, physical properties of the displays and controls, capabilities for recovery from error before significant effects occur, regimens of training and motivation, management policies, and so forth. The performance-shaping factors are then used to "tune" the fault- and event-tree analyses (Embry, 1976).

Markov Network Analysis

Accidents typically occur not because of a single failure but because of multiple failures in combination: A driver ran a red light, another car was entering the intersection on green light, there was a patch of ice, and nether driver saw the other. Given this fact of combination, and the fact that serious failures cannot occur until multiple variables go into failure state, it is promising to analyze failures on a *Markov probability network* or *safety state network,* such as shown in Figure 4.7. Each block in the diagram represents a *safety state:* a vector of three variables (in this extremely simple case, the speed, track curvature, and wheel integrity of a train, each of which can be either in normal or in failure mode; think of it as a three-bit computer word). One can examine the probabilities of changing from any one state to another.

Data tend to be readily available for the probabilities of change from no failure to single failure, but the multiple-failure situations that lead to severe accidents are rare, and no data are directly available. However, one can infer from the single-failure data how the multiple failures might occur and how the failure state might gravitate down the network (Lanzilotta &

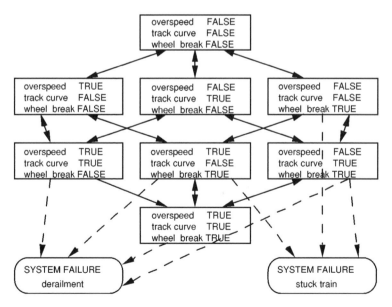

Figure 4.7 Example of safety state network for failure, no-failure of three-element safety vector.

Sheridan, 1997). For this kind of model, *Monte Carlo simulations* are a good way to estimate probability transitions to multiple failure states. In a Monte Carlo simulation, on each of many trials the binary state for each component is set according to a "biased coin flip," with the bias determined by the given probability. After large numbers of trials the end state probabilities are determined by the counts over all trials. Degani, Shafto, and Kirlik (1999) used a similar form of state transition network to analyze human performance in using the aircraft flight management system.

Time Continuum Analysis

Some risk investigators (Askren & Regulinski, 1969) believe a more appropriate way to consider system reliability is in terms of *time until failure,* or, if one wishes to use empirical statistics, *mean time between failure* (MTBF). Confidence limits may be established for the MTBF based typically on a skewed distribution of times until failure, such as the log-normal model. Assuming the human operator is good at recovery or repairing, one can incorporate data on mean time to repair (MTTR) and generate statistics for the fraction of time a given system is available or operational.

Chapter 5

DISPLAYS AND DECISION AIDS

In the present context, the term *display* refers to the presentation of computer-processed information to any of the human senses, not necessarily just to the eyes (and I am not talking about store window displays, Christmas decorations, or magazine advertisements). The term *decision aid* here refers to computer processing of information to help a person make decisions. (Here I am not interested in paper maps, notebooks, tax consultants, marriage counselors, or other noncomputerized forms of decision assistance.) Because operators make decisions in systems on the basis of what information is displayed to them, and because decision-aiding information must somehow be displayed, making a distinction between the information and the display is not particularly useful.

Display design is a common human factors problem, and I will not review the factors of visibility (e.g., size, brightness, contrast, color, exposure time, accommodation, adaptation, vibration, and masking), depth perception (stereopsis, parallax, shadows, etc.), or visual perception and semantics (icons, expectation, illusions, etc.). These may be found in standard texts on human factors such as those cited in the introduction; also see the artful books by Tufte (1983, 1997). Nor will I discuss corresponding factors for auditory or haptic displays. The emphasis here, rather, is on aspects of displays that are particularly related to computers, decision aids, and automation.

Display Dedication, Integration, and Flexibility

Display Dedication and the "Keyhole Problem"
In former times each display was hardwired or mechanically connected to a unique sensor or sensing amplifier that fed it; thus the more variables to be sensed, the more dials or other displays that had to appear on the control panel. With computers that is no longer necessary. A single computer display can theoretically serve all display functions — but

should it? Should each variable have its dedicated display fixed in a particular location, or should variables share a single or small set of computer displays?

The advantage of dedicated displays is that there is a fixed location cue. The operator can identify each particular variable with its own unique location and look there, like finding something in a bureau drawer. The disadvantage is that when there are many variables, there will be a very large number of dedicated displays, and console real estate may get scarce. Figure 5.1 shows how the number of independent displays in aircraft increased until the advent of the Concorde, after which there were concerted efforts to reduce the number of separate displays.

Having many variables share one or a small number of computer screens has the advantage that it takes less real estate, displays can be sized or shaped in any manner, and they can be called up as needed by using menus or other call-up means. The disadvantage of shared screens is what might be called the *keyhole problem.* It is as though the operator were looking at the world through a keyhole. The location cue is lost. Serious problems can arise if the operator does not happen to remember how to access the needed information by using menus, special codes, or whatever is required.

An example of the keyhole problem is when alarm messages are put in a list, with new alarms appearing as they occur at the top of the list and the old ones are automatically scrolled down. Originally this was thought to be a good idea, a space saver, and a convenient means to combine information

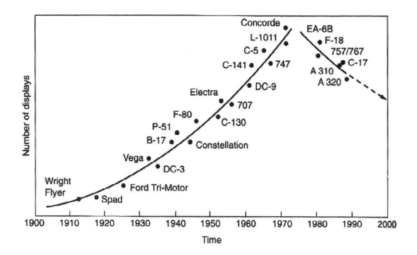

Figure 5.1 Number of separate displays in various aircraft.

on a single screen. However, it was found that if the information is displayed and the operator sees it briefly, then glances elsewhere for a moment and then back again, he or she may find the expected message to be gone (e.g., if some scrolling operation has moved it to elsewhere). That can be very disconcerting. In contrast, the traditional control panel dedicates a particular location to each alarm, so the operator makes use of this location sense in remembering and quickly accessing (scanning) the panel.

The keyhole problem for a single computer screen can be resolved by dedicating at least certain portions of the screen to an overview of information in standard format or to icons with which one can pull up that information, no matter what is being displayed elsewhere on the screen. The Macintosh desktop and Microsoft Windows are based on this principle.

On a large instrument panel, it is not a bad idea to have at least one dedicated display. The safety parameter display system in nuclear power plants presents at a dedicated location a small number (e.g., 6–10) of variables that together give an overview of the "health" of the plant. Even though there may be hundreds of other variables, the operator can see at a glance whether something is abnormal.

Display Integration

Display integration means to combine representations of multiple variables in a meaningful relationship. Computers make display integration possible. For example, in a nuclear power plant both temperature and pressure are important properties of the cooling water at various locations. At an earlier time the plant operator had to examine indicators for both variables and deduce, with the help of a printed steam table, whether the water was in the form of liquid or steam. However, with a computer a simple two-dimensional plot of temperature and pressure can easily represent the combination as a point, and this can be displayed relative to a fixed phase-change line (Figure 5.2), which is often what the operator really wants to know.

Figure 2.1, which showed an aircraft horizontal situation display combining compass heading, predicted course, way points, weather hazards, and relevant navigation fixes and frequencies, is another example of an integrated display.

Integration can also be accomplished with an auditory display. For example, an early experimental Flying by Auditory Reference (FLYBAR) display — designed to relieve the pilot's visual workload in monitoring aircraft bank, speed, and pitch — consisted of a tone that swept in loudness from the pilot's left earphone to the right earphone, in doing so, changing the pitch (low to high, high to low, or constant) to convey as a "sound

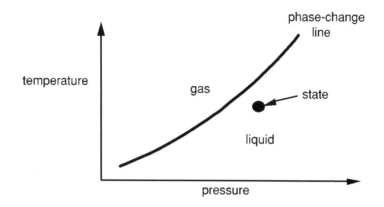

Figure 5.2 Phase plot of gas and liquid.

picture" respectively left-bank, right-bank, or wings level (Chapanis, Garner, & Morgan, 1949). A background "putt-putt" sound indicated speed. Various schemes were tried to convey pitch. Pilots could actually fly in instrument flight rules conditions with this display, but it was never implemented.

Adaptive Display Formats

The computer enables lots of things that may seem like a good idea but may not be. One is the ability to change both the format and the logic of the display as a function of the situation. Most displays in aerospace and industrial systems now have fixed formats (e.g., labels, scales, and ranges are designed into the display). Alarms tend to have fixed set points. However, there is the potential for future computer-generated formats to be reset automatically at various mission stages or in various conditions. Thus for a given aircraft display, the format of a variable could be changed for takeoff, landing, and en route travel, and a process plant display format could be different for plant startup, full-capacity operation, and emergency shutdown. Displays could also be formatted or adjusted to the personal desires of the operator, to provide any time scale or degree of resolution desired at the time.

Adaptive formatting presents problems, however. The operator can become confused by the changing format or may perceive the meaning of a given display in terms of a remembered format that is different from what is currently operative. If there is a shift change of operator or crew, there may be confusion regarding which format any display is currently in. These are reasons to keep the display format fixed. There is particular danger in allowing emergency displays to change format, because they are by definition seldom used and need to be understandable with minimum time and confusion. Adaptive formatting should be done only with great care.

Under changing format conditions some alarms may have no meaning, or they may be expected to activate inadvertently. If adaptive formatting is used, some alarms may be suppressed or the set point may have to be changed automatically to correspond to the operating mode. Adaptive formatting requires special care in ensuring that included alarms and warnings make sense.

Ecological Displays, Compatibility, Animation, and Virtual Reality

Ecological Displays

Following the perspective of the philosopher Heidegger, Gibson (1979) asserted that perception is the acquisition of information that supports action, especially with regard to constraints on action (Gibson called these constraints *affordances*). In his view, actions affect the environment, and the environment in turn affects the action in complete reciprocity. The animal perceives its natural environment with respect to the constraints that determine its action: whether to flee or fight a predator, whether to eat, sleep, hunt, or mate. One school of cognitive psychology espouses these ideas under the rubric of *naturalistic decision making*, stating that humans do not make decisions as normative models would suggest, carefully weighing alternatives and selecting the one with the greatest expected return. Rather, these scientists claim, decisions are *recognition primed* (Zsambok & Klein, 1997). Rasmussen and Vicente (1992) and Christoffersen, Hunter, and Vicente (1997) applied these notions to the design of human interfaces: Displays should be designed so that the constraints are evident and convey in a natural way what is the appropriate action. They called them *ecological displays*.

Figure 5.3 shows an example of one of Vicente's (Christoffersen et al., 1997; Rasmussen & Vicente, 1992) ecological displays applicable to a process control plant, in this case showing inflow (two components) and outflow to a tank, as well as level of fluid. The geometrical configuration makes apparent the constraining relation, namely that

$$Level = time\ integral\ of\ (inflow - outflow).$$

The observer can readily see by the angle of the line connecting the inflow

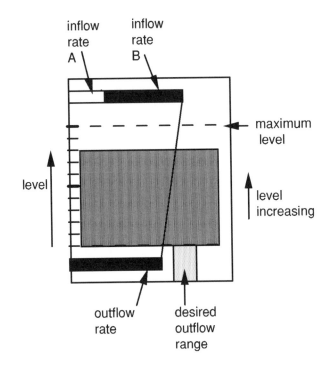

Figure 5.3 Simplification of Vicente's (Christoffersen et al., 1997; Rasmussen & Vicente, 1992) "ecological display" for process control plants. (Adapted from Vicente, 1999, p. 320, with permission.)

and outflow bars whether the level is rising or falling and at what rate. This display can also be said to be "integrated" in the sense described in the previous section.

For a nuclear plant, an interesting experimental display is one that shows the various phases of the Rankine cycle as well as the saturated water and saturated steam lines on a temperature-versus-entropy plot. Although display integration of these abstract elements would mean little to a layperson, for knowledgeable operators it proved better than the conventional set of single-variable indicators found in nuclear control rooms (Vicente et al., 1996).

Compatibility

An accepted principle in display design called *stimulus-response compatibility* mandates that the location of a control (e.g., one of a set of similar-looking controls) correspond to the location of the display (one of a set of similar-looking displays) and that the direction of operation of the control corresponds to the desired direction of operation of the display.

A second principle is that the spatial or functional relationships of adjacent controls and those of the graphical elements within a display should mimic, insofar as it is feasible, the physical reality of the relationships in the real world. For example, the control board layout may correspond to the layout of affected machines or subsystems in the plant.

A third accepted principle is that the direction of operation of both control and display correspond to the *population stereotype*; for example, a variable should increase up or to the right, a standard sequence of operations should go from left to right, and so on. The designer must be careful because population stereotypes differ from one culture to the next; for example, reading is not from left to right in many countries.

The mental model that is taught to the operator should also correspond. The need for this four-way (control-display-physical reality-stereotype/ mental model) isomorphism may be obvious, but it is not always adhered to by the designer, and the consequences can lead to confusion and worse.

A special problem of compatibility is encountered in teleoperation systems in which there is a delay in feedback. Video feedback has by far the highest bandwidth requirement as compared with force or sound feedback, and communication of video signals is likely to be delayed significantly (e.g., 0.6 seconds over an integrated services digital network [ISDN] telephone line because of modulation /demodulation). An experiment by Thompson, Ottensmeyer, and Sheridan (1999) in a telesurgery application showed that under such circumstances, the best telemanipulation performance is obtained when force information is fed back as quickly as possible, even though it leads the video by a full half-second!

Use of Animation and Virtual Reality

Modern computer graphics make it possible for displays to go well beyond what was known in the past, including a variety of animation and virtual reality (VR) elements — much as lecturers making PowerPoint presentations do now. The question is, when should the designer take advantage of this new capability?

Certainly in computer-aided instruction it may be beneficial to show the architecture and workings of a system through animation or VR. Indeed, sometimes that is useful to give the operator a different perspective. For example, one express courier service used VR techniques to have the trainee "become" a package and experience all the rough handling, transportation, storage, and general beating that a package took throughout its whole journey in trucks, airplanes, and warehouses.

Often an operator of a complex system could be helped in making a diagnosis by reviewing how a system works: the physical configuration of an electrical or mechanical system, or the path of signal flow or mechanical forcing. If the operator could ask for a demonstration (model-based simulation) of a given system under various constraints that he or she sets as parameters, on a time scale that he or she prescribes, this might aid considerably in understanding a problem, diagnosing a failure, or planning a control strategy.

However, as noted earlier, just because some technology can be implemented does not mean that implementing it is the best and safest thing to do. The 14th-century principle of William of Occam (Occam's razor), which dictates that simpler is better, is always worth considering. Tufte (1983, 1997) made this point clearly and coined the term *chart junk* for extraneous and superfluous color, shading, or other graphical gimmicks that contribute nothing useful and only obscure the essential message in the display.

Dependence of Decision and Action on How Information Is Displayed

Sometimes displaying the same information in a different way results in a different human response. In an aircraft simulator evaluation of advanced terrain displays, Kuchar and Hansman (1993) demonstrated this fact. One display was a plan view, one a profile view, and one a perspective view (Figure 5.4). When mountains were encountered, the participants' avoidance maneuvers were different among the displays. Flying a given course, the participants initiated lateral maneuvers 80% of the time using the plan view but only 5% of the time with the profile view (and 30% with the perspective view, a compromise). Apparently the avoidance maneuver was chosen based on the ability to confirm avoidance. The plan view was preferred because it provided the most natural confirmation that the turn would be effective. Clearly, how any given information is displayed affects what control action is taken.

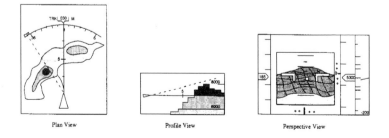

Plan View   Profile View   Perspective View

Figure 5.4 Different aircraft displays of the same information that produce different pilot responses (Kuchar & Hansman, 1993).

Interaction through Icon and Menu Displays, Voice, and Haptic Inputs

There has been much development of interactive tools for computer workstations for word processing, spreadsheets, drawing, and computer-aided design. Icons and menus have become standard on computer screen desktops. Voice and haptic inputs (including touch screens, force-feedback handles and mice, and video and electromagnetic gesture recognition) are now commonplace.

Icons and Menus

Icons are easily learned and quickly recognized. However, caution must be used to ensure that they are clearly distinguishable and there are not too many of them. As with highway signs, medicine packaging, and other displays in everyday life where confusion has serious consequences, it is advisable that computer display icons differ from one another in more than one attribute. For example, they might be made to differ in size, shape, color, and other graphic features. Adding small alphanumeric labels to icons is not a bad idea or, at least, giving the operator the option to ask for such a label as needed, as is done in word-processing software.

Menus have also worked well in computer applications, but, again, it is common experience that they can become too complex. The Windows Start menu selection with multiple hierarchical levels can be very confusing to the beginner. In applications in which speed and safety are critical, errors in menu selection can have dire consequences. It is recommended that menu hierarchies not be more than three levels deep. Mitchell and Saisi (1987) experimented with qualitative graphic icons and adaptive windows as workstation interface aids, and they showed how they improved performance in ground-based monitoring and reconfiguration of Earth satellites.

Auditory menus are commonly used by telephone service providers, but these menus are constrained because there are limits in the user's immediate memory for the items. Auditory menus should number no more than four per selection (level in the hierarchy). After three or four selections, the user can become frustrated, especially if there are branching hierarchies and he or she is expecting to talk to a live person.

Voice

IBM, Dragon, and other software manufacturers have produced speech recognition software now in common use in personal computers, and telephone menus now also make use of speech recognition software. The major problems that have plagued speech understanding research for many years are (a) understanding connected speech (without pauses between words) in context and (b) word recognition independent of the speaker or variations in manner of speaking for the same speaker. Speech recognition software requires significant training by a given speaker.

Although earlier research (e.g., Mitchell & Forren, 1987) suggested that augmenting keyboard commands by voice (recognition) was not a good idea in supervisory control at the time, intent recognition and context-dependent decision making are now much further advanced. At this writing the Massachusetts Institute of Technology Artificial Intelligence Laboratory has a publicly available system called Jupiter (1-888-573-8255), which allows the user (with no training session) to ask for the weather anywhere in the world; a speech recognizer processes the request, searches the Web in real time, and provides a synthesized speech answer.

Haptics

Virtual force feedback was introduced into the hand controls of flight simulators (Repperger, Phillips, & Chelette, 1995) and, later, into the master arms for controlling remote manipulators for handling radioactive materials (Sheridan, 1992b). More recently, virtual haptic interface devices such as the three-degree-of-freedom (DOF) Phantom arm or the Immersion force-feedback 2 DOF mouse permit the user to interact with display graphics and menus by touch (Durlach & Mavor, 1995). Both of the latter have high-resolution displacement transducers and frequency response of several hundred hertz, so the user can "feel" what is seen on the display screen by hard edges, texture of surfaces, and so forth.

One practical use of haptics is for virtual switches and multiposition knobs that can be seen on the visual display and also "felt" when they are manipulated; they might even be heard through an auditory display (e.g., a click). Another use is for differential force or vibration cues to identify position within a menu list. A third use is for training simulators for teaching surgeons in combined visual-haptic manipulation tasks such as endoscopic or arthroscopic surgery.

Alarms and Warnings, and the Problem of Monitoring for Abnormalities

As mentioned earlier, in traditional control rooms and cockpits, a common design practice has been to provide the human operator with independent displays of key variables from every subsystem and for a large fraction of these to be accompanied by alarms or warning displays that light up when the corresponding variables reach or exceed some preset values. This practice has resulted in part from the fact that different subsystems are designed and manufactured by different companies, and this is a way for each company to protect itself from accident liability claims that full information to operate its equipment was not provided.

Complexity

A modern airplane may have more than 1000 displays (including separate computer display pages), and a modern chemical or power plant may have more than 5000. Monitoring for abnormality becomes a travesty when the number of alarms is this excessive. In my experience in one nuclear plant training simulator, during the first minute of a loss-of-coolant accident, 500 displays were recorded to have changed in a significant way, and in the second minute there were 800 more.

Clearly, no operator or team of several operators could be expected to know all of what occurred on that panel. Recognition of this fact has motivated rethinking of display systems for human control of complex processes, particularly with regard to information overload. The availability of computer generation of displays has not automatically eased this problem, and it could even exacerbate the mental workload. The technology allows information to be packed up to almost any density onto a single screen. There is a kind of technology imperative — "If the information is available electronically, make it available on the display" — that must be resisted.

Mumaw, Roth, Vicente, and Burns (2000) emphasized that what makes monitoring of complex displays difficult is not the identification of subtle indications against a quiescent background but, rather, the identification and pursuit of relevant findings against a noisy background. They also asserted that operators need to be proactive problem solvers.

Adelman, Cohen, Bresnick, Chinnis, and Laskey (1993) reported a study that suggested it is not a good idea to force the allocation of an operator's attention, that an interface should highlight what it (the automation) thinks warrants attention but nevertheless allow the operator freedom in allocation of attention.

Intercorrelation and Inhibition of Alarms

The multiple alarms and warnings displayed for almost any real operating system tend to be highly correlated. Clearly, many different pieces of equipment tap into the same electrical power buses or pipes, and differences in temperatures, pressures, flows, or voltages from one point to another often are predictable functions of one another. In other words, the different variables tend to be coupled physically and causally.

Experienced operators claim that they perceive patterns in the alarms and warnings that correspond to various situations, but these operators don't pretend to understand how they do it. They affirm that it takes considerable experience and effort. This is true whether displays are *analogic* (e.g., diagrams, plots), *symbolic* (e.g., alphanumeric), or some combination. The challenge is to design displays to integrate the information and thus enable human operators to perceive patterns in time and space reliably and without great effort or experience. Although some redundancy is helpful, there can easily be too much, to the point that the operator is wasting time observing the same information from multiple sources and missing other information that is critical.

One solution for this excess redundancy is to display only the variable or variables at the point of failure, assuming that point can be isolated, and to actively inhibit displays of variables that are causally downstream from the failure point, which would necessarily be triggered even though the initial failure is upstream. The problem is that minor abnormalities at one location can compound and lead to significant failures at other locations, so isolation of the root cause of a problem is not always easy. For any two displays that are correlated most of the time, there may be times when they are not correlated, and the operator ultimately should have access to both displays. However, this poses the redundancy dilemma. Ultimately it is a complex problem in alarm decision logic, potentially a job for the computer to sort out if enough human expertise can also be put into the rule base.

Nuisance Alarms and Warnings

Every alarm or warning system is set at some threshold for the level of the variable or combination of variables. If the threshold is too low, every potential hazard will trigger that alarm, but so will situations that are not considered hazardous (so-called false positives or false alarms). If the threshold is set too high, many situations that should be alarmed will not be (so-called false negatives or misses). A balance or compromise must be struck. How to decide on appropriate alerting thresholds is an active area of research (Kuchar, 1996).

It is also commonplace in complex systems for parts of the system to be checked, maintained, or repaired while the system is operating, and alarms and warnings associated with that equipment may be inoperative in either an on or an off state. Other alarm/warning circuits may themselves be malfunctioning, even though the equipment they monitor is functioning properly. Ideally, the system would not be allowed to operate unless everything was working properly. Unfortunately, that is not realistic. This leaves the human operator with the task of keeping track of which subsystems are out of service and which alarm/warning systems are malfunctioning — or, perhaps worse, suspected of malfunctioning, without convincing evidence.

Displays to Aid with Monitoring and Procedures

For the most part, controlling an automatic or semiautomatic system is a monitoring task. Most of the time alarms and warnings are not going off, and monitoring can be boring. The result of this mental underload can be inattention, degradation of vigilance, and drowsiness, as noted earlier (Chapter 4). It has long been established that after about 30 min of monitoring, an operator can easily become unalert (Mackworth, 1948). Means of maintaining vigilance are badly needed (Molloy & Parasuraman, 1996). A fact of life is that human beings are not very good monitors. Still, as systems become more automated, and insofar as it is advisable to have humans involved somehow, monitoring is what humans must do.

Part of monitoring is to maintain a running correspondence between the system state and the operating procedures. Normal operating procedures are mostly committed to memory by operators in sophisticated systems such as aircraft, nuclear power plants, and robotized manufacturing plants. However, there are always questions of whether operator memory is sufficient, especially if an abnormality is suspected.

Traditionally, operators have been given paper checklists and procedures documents and then assigned to step through these, log their readings and other observations, and check off the completion of each step. Although some may regard this as busywork, it does tend to keep the operator alert and in the loop. Can the computer play a useful role in such monitoring and procedural activity and in helping to maintain operator alertness?

Computer-based "smart checklists" are already being developed for aircraft, in which the computer checks to see whether the operator really took the action called for in the procedure and, if not, calls attention to the discrepancy. Other computerized systems have been developed that display to the operator, based on the system's own estimation of past and present process states, the procedures that best fit that plant condition (e.g., what to do in response to an alarm, or what to do at a particular time or stage in the procedure). In this case procedures are not implemented automatically, though in some cases they could be. Usually it is deemed important to give the human operator some room for interpretation in light of his or her knowledge about the process or the current operating objectives that the computer does not understand.

In the Office Systems section of Chapter 2, *computer-supported cooperative work* was mentioned, referring to networked computers serving to integrate communities of people in problem solving, or teams of people operating in geographically distributed complex systems. Mostly these developments have occurred for business systems. More and more such systems are being developed for flight or rail transport operations, satellite control from the ground, and power distribution (Jones & Jasek, 1997; Mitchell, 1987).

Model-Based Predictor Displays to Aid with Control

There can be trouble whenever there is a time delay in feedback from the controlled process. That is because within any closed loop, positive feedback results when there is energy at a frequency such that the closed-loop time delay causes a 180° phase shift. Positive feedback with any loop gain greater than one means that energy is continuously amplified. This essentially defines instability (see Appendix A).

Time delays are of two types. The first is the discrete delay of a tele-communication signal: After the delay, the signal is reproduced without distortion. Telecommunication delays occur because of the limited speed of electromagnetic waves in space communication (300,000 km/s), limited speed of pressure waves in acoustic communication (speed differs with the density of the medium), and limited speed of serial data processing (e.g., modulation/demodulation of standard video over ISDN telephone lines may take 0.5 seconds).

The second type of delay is the dynamic filter or lagging delay, which is really a distortion in which signal components at different frequencies experience different reductions. This is attributable primarily to mechanical viscosity-inertia and thermal-hydraulic equivalents. These dynamics occur in nuclear, chemical, and other process plants and in vehicles such as large ships or aircraft.

Predictor for Discrete Delay

Figure 5.5 shows how a predictor display system works in the case of discrete delay for space telemanipulation. Using the human operator's control inputs, a computer simulates the kinematics without delay and immediately displays graphically the (simulated) system output, usually superposed on the display of delayed video feedback from the actual system output. In effect, insofar as the simulation is faithful to reality, the model-based simulation is a prediction of what will be seen after the time delay. However, disturbances may exist in the actual environment that are impossible to anticipate or model that will make the actual remote system deviate from the model. Even with small disturbances and imperfect models, however, such a technique has been shown to allow a 100% increase in the speed of remote task completion (Sheridan, 1992b).

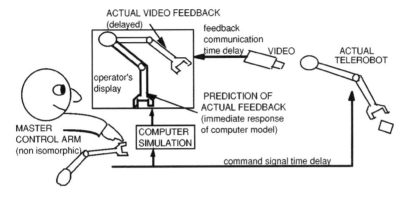

Figure 5.5 Predictor display system for discrete communication time delay in space telemanipulation.

Predictor for Dynamic Lag

When the time delay is of this type, the scheme for predicting is somewhat different. The computer simulation is dynamic rather than merely kinematic or otherwise static.

Figure 5.6 tells the story. Consider a large ship, and consider first-order dynamic approximations: Ship lateral position y relative to desired course is the time integral of ship heading angle deviation α times forward speed. Heading angle deviation is the time integral of rudder angle φ times forward speed. Rudder angle is the time integral of command signal z times motor drive speed. This is modeled in the computer by a cascade of three integrators (bottom row of blocks) with time constants that match those of the actual ship and thus are fairly long, probably minutes.

The top row of blocks in Figure 5.6 shows a corresponding series of three integrators with time constants set to simulate what the real ship

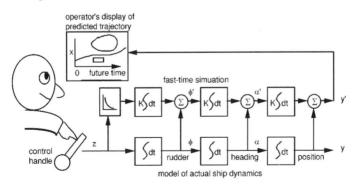

Figure 5.6 Predictor display system for long-lag ship dynamics.

does, only much faster, perhaps 100 to 1000 times faster. This simulation is repetitively reset to zero, and the current values of the (very slowly changing, essentially constant) initial conditions for y, α, and ϕ are added as shown. Based on the current input z, on each new prediction cycle, the fast simulation extrapolates ahead from where the actual ship is now. Note further that there is a discounting block that weights the current z value heavily in the first phase of the fast simulation and less at the later stage of the simulation cycle. This assumes that the future z is undetermined and therefore discounted eventually to zero (i.e., the operator is not expected to hold the control at a constant value). The resulting, continuously changing (as the control handle is slowly changed) predicted trajectory is displayed for the operator, relative to the desired course and the obstacles.

Decision Aids for Satisficing

Decision making often means selecting among a large number of alternatives, none of which is dominated by any other alternative. (To be *dominated* means it is worse with respect to at least one objective or attribute and no better in any other; see Appendix A, section 4.1). The set of such nondominated alternatives is called the *Pareto frontier* in the space of multiple objectives or attributes that together define relative importance or worth. Formally and ideally, the unique point on the Pareto frontier with the greatest utility is the optimum (see Appendix A for a formal explanation). However, as noted earlier, real decision makers do not seem to execute all the steps of normative decision optimization; at least, those steps are not overt in their behavior. Real decision makers *satisfice* — that is, they narrow down to a set of acceptable alternatives or a region in continuous decision space that is "good enough," such that further decision effort is not warranted. Also, the more familiar the context, the more quickly and easily real decision makers are able to satisfice. Such decision making often needs little help from the computer.

However, some important decisions cannot be made so easily and quickly, because the consequences are large and/or because the alternatives have many attributes and the decision space is simply too large for the decision maker to fully comprehend the interrelationships. In this case the computer can help, though the decision cannot be left wholly to the computer. The reason is that the *utility function* (or *objective function*), the trade-off between all the salient objectives or attributes as they apply to current circumstances, is something that is uniquely human and something computers may never be able to cope with fully, essentially by definition of what a computer is.

How can the computer help? By doing what computers do best and interacting with the human to let the human do what he or she does best.

To convey the idea of how a computer can aid in satisficing, consider Figure 5.7, which depicts a decision space of only two objectives (ride quality and speed) for comparing alternative means of transportation. Making a decision in a two-objective space would normally be so simple that the computer is not necessary. However, if the reader can imagine a hyperdimensional decision space of, say, 10 or more objectives (which is impossible to draw), the point will be made.

Suppose there are many possibilities to choose from but that the set of possibilities is constrained by physical reality — as speed goes up, ride quality necessarily comes down (the straight line on the left of Figure 5.7). Also suppose there are problems of manufacturing that constrain speed and ride quality, which together might determine a Pareto frontier bounded by the straight line at the right. Suppose further that there is an imposed budget, which adds another constraint to the Pareto frontier specified by the curved line. For a large multidimensional decision space it would be easy for a computer to keep track of which points lie in the feasible side of the frontier (down and to the left of the lines in this example) but impossible for a human.

Now suppose that the human decision maker expresses the hope that Point 1 is feasible. The computer can quickly tell him or her that it is not possible but can suggest that the closest point to the Pareto frontier is Point 2. At that stage the decision maker might think, "I really want better ride

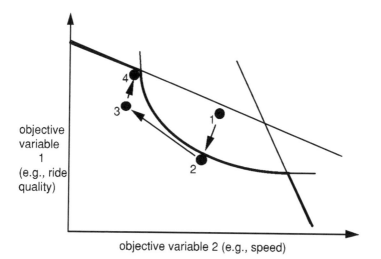

objective variable 1 (e.g., ride quality)

objective variable 2 (e.g., speed)

Figure 5.7 Display of human-computer interaction for satisficing in two objectives.

quality and would be willing to sacrifice a little speed, so let's try Point 3."
Then the computer would come back and point out that the decision maker
can actually do better in both objectives by Point 4. In this way the
computer and human can continue to take turns and interact, until the
human finally thinks, "OK, this is close enough" (a satisficing point) and
quits the procedure.

Charny and Sheridan (1989) constructed such a system with a graphic
user interface for five-dimensional spaces, and they found that users had
little trouble making satisficing decisions that might otherwise have been
infeasible or significantly inferior to what they could have made.

Etiquette of Conversing with a Decision Aid

Etiquette is not a term normally employed by system design engineers.
However, as machines become more intelligent, and as it is desired for
such machines and the people who use them to communicate and cooperate, it is evident that the same social mores that apply to people also
apply to intelligent machines. What humans consider to be good etiquette
is a set of behavioral practices that make interaction between mature
people acceptable and efficient (Grice, 1975). If it were not so, etiquette
would have died out long ago. Grice posed four maxims for cooperation in
conversation:

1. Maxim of quantity: Say what serves the present purpose but not more.
2. Maxim of quality: Say what you know to be true based on sufficient evidence.
3. Maxim of relation: Be relevant, to advance the current conversation.
4. Maxim of manner: Avoid obscurity of expression, wordiness, ambiguity, and
    disorder.

Miller (2000) discussed the rules of etiquette as might be applied to
adaptive user interfaces. His proposed rules, slightly abbreviated, are as
follows:

1. Make many conversational moves for every error made.
2. Make it very easy to override and correct any errors.
3. Know when you are wrong, mostly by letting the human tell you.
4. Don't make the same mistake twice.
5. Don't show off. Just because you can do something does not mean you
    should.
6. Talk explicitly about what you are doing and why. (Your human counterparts
    spend a lot of time in such meta-communication.)
7. Use multiple modalities and information channels redundantly.
8. Don't assume every user is the same; be sensitive and adapt to individual,
    cultural, social, and contextual differences.

9. Be aware what your user knows, especially what you just conveyed (i.e., don't repeat yourself).
10. Be cute only to the extent that it furthers your conversational goals.

It would seem that Grice's (1975) and Miller's (2000) ideas are relevant for making decision aids more useful and acceptable.

The Ultimate Decision Aid

Before ending this chapter on display and decision-aiding aspects of humans and automation, I should note that various researchers have contemplated what a so-called ultimate decision aid might be like. The U.S. Department of Defense has sponsored projects pointed in that direction, in one case a computer-based "pilot's associate" to assist fighter pilots, and in another case a command and control decision-aiding system for commanders of naval ships.

A human assistant — say, a copilot in an aircraft — must maintain a current mental model of the person to be aided (in this case, the pilot flying): what he or she has done, is doing, intends to do, and can be assumed to understand about the current situation. The assistant must also have situation awareness of the system being controlled and its environment. This is a tall order, necessitating maturity and sophistication with respect to knowing how the system works, its mission, the physical and other constraints of performing the task, and so on. It is useful to think about decision aids in terms of what a human assistant must know, assume, observe, decide, and communicate to the person being assisted.

Figure 5.8 illustrates the complexity of the situation in terms of controlling a robot. The human operator (like the pilot flying in the previous example) is normally expected to have some mental model (the "thought balloon" in the figure) of the robot and the task and to have some understanding of the decision aid: under what circumstances is it useful, what information it can provide, and how trustworthy is it. However, such *meta-knowledge* is not typically expected of decision aids. The decision aid of the future should have an internal model not only of the robot and task but also of the human operator, as represented in the thought box above the decision aid.

Rule-based expert systems are hardly a new concept; they were first developed by artificial intelligence researchers more than 40 years ago. One of the first serious efforts was to provide a rule-based computer adviser to aid physicians wanting to prescribe -mycin drugs (those being the set of miracle drugs of the era). The physician was to tell the expert system what it wanted to know about the patient, and the expert system

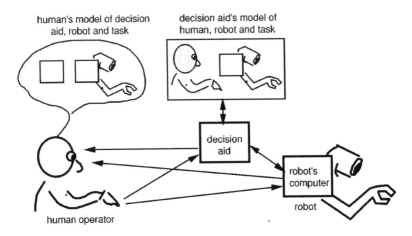

Figure 5.8 The ultimate decision aid: the intelligent computer assistant (shown for a tele-manipulation task).

was to advise which drug to use. Unfortunately, after all these years, physicians have not adopted this or any other expert system for routine diagnosis or therapy. However, rule-based expert systems of both the crisp and fuzzy variety (see Appendix A) have been applied to many simpler situations. Such work is definitely progressing. But as with speech under-standing, language translation, intelligent robots, and many other applications for which the artificial intelligence community was to have solved the problem within a decade, there remains much work to do.

Chapter 6

SUPERVISORY CONTROL

Much of the book up to this point has implied that the human operator engages in a supervisory form of control over the automation for which he or she is responsible. Supervisory control is inherent in human-automation systems. To talk in terms of supervisory control is really a way of generalizing what the human is being called on to do when interacting with automation. This chapter defines and details what supervisory control is and when it is appropriate.

Definition of Supervisory Control

The term *supervisory control* derives from the close analogy between a supervisor's interaction with subordinate people in a human organization and a person's interaction with intelligent automated subsystems. A supervisor of people gives directives that are understood and translated into detailed actions by those human subordinates. The subordinates, in turn, gather detailed information about results and present it in summary form to the supervisor, who must then infer the state of the system and make decisions for further action. The intelligence of the subordinates determines how involved their supervisor becomes and how often he or she needs to interact with them. The same sort of interaction occurs between a human supervisor and the automation (Sheridan, 1992b). The automation is a subordinate with limited intelligence.

In a strict meaning, *supervisory control* means that one or more human operators are intermittently programming and continually receiving information from a computer that itself closes an autonomous control loop from the artificial sensors and through the actuators to the controlled process or task environment. In a more generic, or liberal, meaning, *supervisory control* means that one or more human operators are intermittently programming and receiving information from a computer that interconnects through artificial sensors and effectors to the controlled process or task environment.

Is there an essential difference between these definitions? By both definitions, the computer transforms information from human to controlled process and from controlled process to human. It is only under the strict definition, however, that the computer necessarily closes a control loop that excludes the human for at least some of the time, thus making the computer an autonomous controller for some variables for that period. The time span of computer control can differ for different variables being controlled by the same human and can differ from one such period to the next for the same variable — depending either on what the human specifies or on what the computer is given authority to decide.

The strict and generic forms of supervisory control may appear similar to the casual observer, given that the supervisor always sees and acts through the computer (analogous to executive staff). The casual observer may not be aware how, at lower levels (e.g., for individual machine operations on the factory floor), the computer is actually implementing supervision. If it is the generic type, the computer is acting in an open-loop fashion, integrating and interpreting incoming information for presentation to the human decision maker, or executing a specified string of commands set by the human programmer. If it is the strict type, the computer is working in closed-loop fashion, making decisions based on what it senses, then acting on those decisions.

In the strict form of supervisory control, the human operator (supervisor) programs by specifying to the computer goals, objective trade-offs, physical constraints, models, plans, and "if-then-else" rules. This specification is usually and most conveniently put in high-level "natural" language — in terms of desired relative changes in the controlled process, rather than in terms of low-level control signals. Then the supervisor turns control over to the computer, and the computer executes its stored program and acts on new information from its sensors independently of the human.

The human may remain as a supervisor of computer control throughout a whole operation or, from time to time, may switch to direct manual control (this switching is called *trading control*). At times the supervisor may manually control some variables while the computer simultaneously controls other variables (this is called *shared control*), as was illustrated in Figure 3.5.

The Historical Trend

If one considers what has happened historically as human-automation systems have become more sophisticated, it is a progression of stages as depicted in Figure 6.1: At first (Panel 1) the human controls the process directly, with his or her own eyes and hands. Next (Panel 2), dedicated display and control devices are placed between human and control process, mandating a form of indirect control but enabling the human to be elsewhere in space. Then (Panel 3) the computer intercedes in the loop, enabling both control and display to take on new and improved functions and make the human's job easier. This is the beginning of supervisory control as defined liberally.

At the next stage (Panel 4) there is finally supervisory control by the strict definition given in the previous section — the computer is in charge at least for short periods. Panels 5 and 6 are refinements on Panel 4. In Panel 5 a physical separation develops between the computer that talks to the human (the *human-interactive computer*) and the computer that talks to the controlled process (the *task-interactive computer*) with an electronic communication channel in between (two computers talking to each other). This occurs, for example, in any system in which the control room or flight deck with its computer system is separate from the computers local to various sensors and actuators out in space, down in the ocean, out in the plant, or on the airplane wings. Panel 6 illustrates when one human-

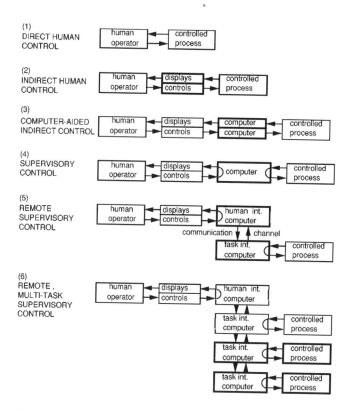

Figure 6.1 Levels of control from purely manual to multivariable and remote supervisory control.

interactive computer controls many task-interactive computers. The last stage is already prevalent in a great variety of complex systems.

It should be noted that supervisory control is not a theory, model, or philosophy of design or operation, although theories, models, and design philosophies for dealing with supervisory control are important and needed. It is, rather, a term referring to a factual role relationship of a human to his or her physical environment, historically a relatively new one, having emerged in conjunction with automation over the short span of the last 50 years.

Five Roles of the Human Supervisor

Five roles can be cited for the human supervisor. The roles are undertaken more or less in causal and temporal order, with looping back and recycling as diagrammed in Figure 6.2.

The first role is for the supervisor to *plan* what needs to be done over some period, before any automation is turned on. The second role is to *teach* (instruct, command, program) the computer with what it needs to know to perform its assigned function for that period. The third role is to turn on the automation and *monitor* the automatic action, detecting any failures. The fourth role for the supervisor is to *intervene* in the automatic action as necessary, deciding on and making any necessary adjustments to

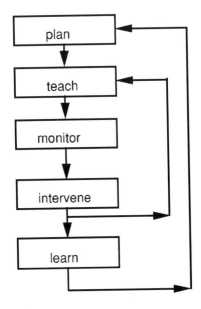

Figure 6.2 The five roles of the human supervisor.

the automation. The fifth role is to evaluate performance and *learn* from experience.

Notice the feedback loops in Figure 6.2. The inner loop is active when the supervisor intervenes and must reteach the computer. The outer loop is active when the supervisor learns from experience and applies that learning to planning the next task.

Planning

First, the supervisor must have knowledge of the physical elements (or mix of people and machines) to be controlled — the variables and the controlled process. The variables are of two kinds: *independent* (input, *exogenous*) and *dependent* (output, *endogenous*) variables. The variables that are independent of the controlled process in turn are of two kinds: First are the independent variables that cannot be affected by the operator but must be expected by him or her; if they are undesirable and must be compensated, they are called *disturbances.* Second are the independent variables that can be manipulated manually by the operator or operators or through the automation — the *control variables.* The dependent variables are typically many, but usually only one or a small set of these are of interest. In combination (a vector) these define the *process state* or *system state.*

The human needs to understand as much as possible the relation between input variables (of all kinds) and the process state. This is what is usually called the operator's *mental model* of the process — in effect, how the process "works" at the most elementary level of understanding: "I do this (or nature does that), and the state variable responds in this way."

Second, there is a deeper level of understanding the process. Dependent variables that are not of interest in normal operation (not state variables) may nevertheless be of great interest in diagnosing system failures or abnormalities, and they therefore become important to the mental model. Thus if the operator seeks not only a functional understanding of normal operation but also a physical understanding of the system, he or she must know the independent-dependent relations for a larger set of variables. Theoretically the number of variables and relationships in that case is without limit. There is a reasonable limit to what can or should be expected of an operator in terms of understanding enough of the physics of what is happening. (The system designer should consider what is reasonable and take this into account in the design.)

In addition to having a mental model of how the process works, the human operator (and the system designer) must understand resource constraints: energy or fuel, time, and money — what is available and what is reasonable to use. The system designer must consider the ergonomic and cognitive limitations of the human operator.

Third, the supervisor must understand the performance goals or objectives and how they are to be traded off against each other and against constraints (see Appendix A for a discussion of the trade-offs between objectives). Human-automation systems cannot be all things to all people, so there must always be some type of trade-off between goodness of performance and cost in resources of various kinds.

Planning is a quintessential human function, so it is unavoidably left to the human supervisor and, before the fact, to the designer. Planning includes integrating the three types of understanding just mentioned, even though the elements are typically not stated quantitatively. However, human operators, especially those with experience, have a remarkable capability to put the three elements together to perform what is called *satisficing:* coming up with plans or control strategies that are not globally optimal but that are acceptable or satisfactory (Charny & Sheridan, 1989; March & Simon, 1958).

Information acquisition and interpretation for planning of control is what people are especially good for in systems. Nevertheless, people need not do all the planning by themselves. Computers can store data based on past experience, fuse data from current measurements, apply a large number of crisp or fuzzy rules from a preprogrammed rule base, run "what would happen if" simulations, and display the results to the human in whatever format is desired, including fancy computer graphics and virtual reality. Call this decision aiding or automation of information acquisition. These new aids for acquiring information and planning pose many new problems of how much is too much. Decision aiding was discussed in Chapter 5, and more will be said in Chapter 8.

Teaching

Given the plan, the supervisor must decide what to tell the computer to do and how to communicate those instructions. These are two separate tasks; neither is trivial, and both are common sources of error. The first requires translation of the plan into functional steps, and the second requires the translation of the functional steps into some formal command language that the computer can understand.

Command languages can take many forms: dedicated analog controls, such as control sticks and knobs; dedicated symbolic controls, such as discrete push-buttons; concatenated key presses of alphanumeric characters to form computer code; interaction with computer graphic touch screens; and speech in a limited vocabulary that is recognized by the computer.

Experimental studies by Brooks (1979) and Yoerger (1982) in teaching a robot to perform a task clearly showed that giving symbolic commands *relative* to given environmental objects and/or relative to some reference task or template (i.e., teaching the *differences*) is more efficient than demonstrating the entire desired movement in analogic fashion.

For the most part, command languages or programming should follow standard structures and intentionally not be too flexible (it was noted earlier that too much flexibility can lead to errors). As with desktop computer applications, it may well be desirable to have menu-driven commands available as well as shortcut key commands for those who are more experienced or can be expected to remember them.

Different procedures apply to different phases of the task and different contingencies, so the human operator must either have these memorized or have ready access to information reminding him or her of the procedures.

A serious problem occurs when the computer has an inadequate capability to decide if it is getting appropriate commands for the current situation and to give feedback to the human supervisor about what it understands it has been commanded to do. This is a different kind of feedback from the feedback of action the system has taken. In normal discourse between two humans, there are subtle cues by which the speaker gets to know if the listener has understood correctly, and/or the listener can ask for clarification. Figure 6.3 illustrates the point. One would like to enhance supervisory command language interfaces with such a capability. Some business software already does this in a modest way.

Monitoring (Information Acquisition and Analysis)

When the automation is turned on, the human becomes a monitor, acquiring information about process state. *Monitoring* means continuously (or continually) observing many different variables, trying to be aware of

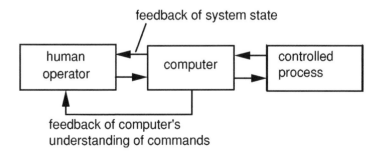

Figure 6.3 Feedback of computer's understanding of commands.

the general situation, and being on the lookout for abnormalities or failures. Direct actions on the controlled process that do not amount to parameter or program changes to the automation are considered part of the monitoring function.

By what strategies should the human supervisor share attention among many variables, assuming it is impossible to attend to more than one at a time? Operations can be performed in parallel by essentially different computers, and there is plenty of time for a single computer to time share. The same cannot be said about the human operator, however. The human, relatively, is very slow and cannot shift attention rapidly from one function to another.

It might be expected that the operator would rotate attention in a regular round-robin pattern among the salient variables. However, that would not be efficient if some variables changed very rapidly and others changed very slowly. Sampling theory dictates that sampling of any variable should be at least once for every half cycle of the highest frequency in that signal, or else something will be missed (see Appendix A). Proper allocation of attention obviously also depends on what is most urgent, what is most important, and what are the required human and physical resources (Wickens, Tsang, & Pierce, 1985).

Tulga and Sheridan (1980) ran experiments in multitask monitoring and compared participants' decision behavior with what was optimal. Results indicated that the participants' choices were near-optimal given ample time to plan ahead. As the task demands became heavier and the mental workload grew, however, the strategy switched from "carefully plan ahead" to "do whatever is most urgent." Interestingly, the participants found that as pace and loading increased and planning ahead finally had to be abandoned in favor of doing what had an immediate deadline, subjective mental workload actually decreased!

Situation awareness is usually defined in terms of missed signals or questions posed to an experimental participant when a simulator exercise is suddenly stopped (Endsley, 1995). In the past operators have had to assess the current situation by keeping track of many individual displays, remembering the past history of these signals, and mentally extrapolating into the future to judge whether or not the trend is bad. Computers now easily integrate the formerly separate and discrete signals into interrelated graphical "pictures" and combine history (trends) and predictions with present status, so as to give overviews with respect to time as well as interrelations between variables. For a given number of variables this makes monitoring easier, but the difficulty is that the human supervisor is being asked to monitor an increasing number of variables.

Much effort has been given to the problem of how a computer might best detect failures and get the operator's attention. There are many techniques. Bayesian updating, as explained in Appendix A, is one method for assigning likelihood to various failure hypotheses based on various bits of available evidence. For continuous data, Kalman filters or estimators are useful (Curry & Gai, 1976).

Figure 6.4 illustrates another technique showing how separate computer simulations of the different subsystems can be used to detect failures and locate the source of a failure. Input signals from the actual system (top) are tapped and fed into the subsystem simulations (bottom), and the corresponding outputs of actual and simulated subsystems are compared. Discrepancies (e.g., at A), if they are large enough, trigger alarms that locate the failure and are displayed to the operator. Note that off-normal inputs (e.g., caused by upstream failures at A) to nonfailed (e.g., downstream) subsystems do not cause discrepancies (e.g., at C). Note also that measurements may sometimes not be available (e.g., at the output of B), so in such cases fault location must be done for combined subsystems (here, B and C). Model-based failure detection/location systems such as this one and Kalman filters are dependent on models that are roughly accurate. For variables in abnormal ranges, good models are often not available.

Once the computer has relevant information about a failure, the problem then arises of getting the human operator's attention. Auditory alarms, either speech or tonal signals, are generally regarded as best because they do not depend on where the operator is looking. Visual indications are then good for providing the details. Lights or visual lists or messages should be arranged hierarchically, as noted earlier.

In any automatic alarming system or human alarming strategy, there is the problem of trading off between (a) responses to false alarms with their

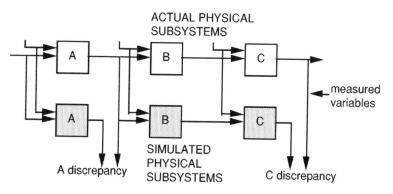

Figure 6.4 Method for using computer simulations of separate subsystems to detect and locate failures.

associated expected costs and (b) ignored true alarms and their expected costs (probability times cost in each case). Because false alarms may be more frequent than true alarms, there is a danger of ignoring all alarms, and the cost of an ignored true alarm is usually far greater than that of responding to a false alarm (Meyer, 2000).

Alarm response is very situation dependent, especially with regard to whether the computer or human should be given the authority to respond to the alarm. For example, Inagaki (1999) analyzed the situation of aborted takeoff of an aircraft, the dependence of the ideal human-computer allocation on when the alarm occurs relative to rotation speed, and the relative trust of the pilot in the alarm. There is a chance that the very mechanism (circuitry) that activates the alarm will be disabled by the events that would normally trigger the alarm (e.g., fire or explosion). There is a need for positive knowledge that the alarm system is working, and if it is disabled for any reason, the alarm itself should signal the operator in some way.

A simple decision analysis of the failure response problem — whether to automate that or not — is given in Appendix A.

Intervention (Decision and Action Implementation)

Intervention occurs when the human supervisor makes any parameter or program change to the automation (task-interactive computer). This occurs when the automation completes a planned operation and it is time to invoke or program the next phase of operation. It also occurs when an abnormality or failure occurs and either the task-interactive computer must be quickly reprogrammed or the supervisor must bypass the automation and manually take over control of the process.

When should the human intervene? This is the phase of supervisory control at which the human operator is deemed most essential because, by definition, if the automation fails, it must be revised or replaced. The decision as to whether and when to intervene is a critical one. On one hand the supervisor, on getting an alarm signal or diagnosing a failure from nonalarmed displays, can react immediately; on the other hand, it may be wise to collect more observations, with hands in pockets, so to speak.

There is an obvious trade-off here: The longer one waits and the more data one gets, the more sure one is of what action to take, but the longer one waits, the worse the potential consequences of not acting. A rational decision depends on the costs in each case, both functions of time. How long an aircraft pilot should wait after hearing the ground proximity warning, "whoop, whoop, pull up, pull up," is likely to be very different from how long an automatic machine tool operator should wait after an indication of tool wear or how long a process plant operator should wait after an indication of low feed water.

Do operators under stress really make decisions? There is a debate about whether human operators under stress make considered, rational decisions, simply react as they have been conditioned from past experience, or do something in between. System designers are quite ignorant here. The stress and confusion are compounded when the alarm or other indications of failure are believed to be unreliable but the consequences of not responding to real failure in time are great. As systems become more complex, with many people and many machines interacting, the confusion escalates. This is the same phenomenon as was earlier referred to as "the fog of war."

What about computer aids to help decide when to intervene and what action to take? As mentioned earlier, computer-based decision aids, which are an inherent part of modern supervisory control systems (part of the human-interactive computer), not only indicate failures but also perform diagnoses as to the cause and recommend remedial actions for the human to take. If such systems were reliable, one could argue that the recommended action should be taken automatically, with the human acting only as a monitor of the automatic takeover (in this case with both automation and human supervision escalating to a higher level). If such advisory systems are somewhat less reliable, however, one must argue that the human should consider the advice and weigh it together with his or her own independent judgment about what to do.

Then the question arises as to whether the human will put too much faith in the computer adviser, and thereby neglect making independent observations and judgments, or will have too little faith, stubbornly believing in his or her own judgment when in fact the computer is both quicker and more often correct. With experienced pilots in a collision avoidance simulation, Pritchett (1999) found a tendency not to follow a new system's recommended avoidance maneuvers, probably attributable in part to disbelief that a real emergency existed, in part to lack of understanding of the system's basis for its recommendations, and in part to habituated responses that were inappropriate in this case. The correct level of trust can be approached by designing the adviser to give not only recommendations but also reasons for the recommendations. This enables the human to make a better calibration of the advisory system.

Computer-based aids to intervention decisions by the operator are rapidly evolving, yet there continues to be a lack of sufficient understanding of human decision making under stress — as well as of the elusive phenomena of trust, situation awareness, and mental workload.

Learning

The fifth role of the human supervisor is that of learning from experience. In order for learning to take place, some measures of performance must be evident.

There is a big role for the computer here, one of measuring performance of both the system and the human. The computer can correlate performance against various criteria set by the human, plot trends, and make recommendations to the supervisor for future plans.

Human supervisor learning and how computers can aid in learning has been an area neglected by designers, but the time is ripe for improvement. The fact that sensors, data analysis, and data storage have become smaller and/or cheaper means that much more can be gleaned from operating experience than in the past. Further, the process of learning is motivating to the operator and may be an antidote to boredom and inattention.

For example, consider driving a car or truck. New sensing and data-processing technology permits the continuous measurement of headways between vehicles, lane keeping, speed, acceleration, and braking. Such data can be used to help drivers evaluate their own performance. This may be especially useful for older drivers posed with the terrible trade-off between ending their days of driving (and all the accompanying problems of inconvenience and loss of self-esteem) and continuing to drive under greater risk to themselves and others.

Anonymous self-reporting of "near misses" in aviation has proven very useful to researchers and designers. Black boxes, which are well-known means of after-accident learning from aircraft crashes, can easily be added to highway vehicles. In aircraft, radar, GPS, data link, and other new sensing technology permit much closer monitoring of both pilot and aircraft performance. Of course, there are concerns for protecting privacy whenever human behavior is measured and recorded.

One must continue to keep in mind that an important reason for having human operators in systems in lieu of complete automation is the ability of the human to perceive patterns and learn. It is usually preferred for human operators to be a bit experimental, not just to follow procedures like mechanized zombies, even if it means making a few small but tolerable errors.

A Composite Picture of Supervisory Control

Figure 6.5 puts the five supervisory roles or functions together (shown above the smiling human operator) and breaks some of these roles into two or three subfunctions (boxes with lowercase labels). Each subfunction box represents not only a cognitive task but also a potential computer aid. To summarize:

1. *Planning* consists of three subfunctions: (a) modeling the physical system, (b) making trade-offs between objectives (*satisficing*), and (c) formulating a strategy for implementation by the automation.
2. *Teaching* consists of two subfunctions: (a) selecting the desired controlled action and (b) teaching (programming) the computer to execute that action. Control is then handed off to the automation through normal computer commands. The various computers execute control and implement their tasks.
3. While the automation is actively controlling, the human is *monitoring,* which consists of three subfunctions: (a) allocating limited attention to various instruments, such as direct observations, communications with team members, and written documents; (b) using the observed data as well as the mental model of

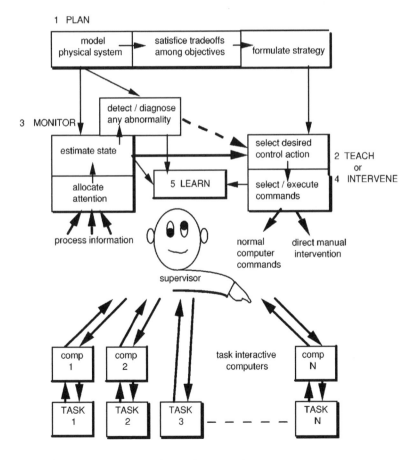

Figure 6.5 A composite picture of supervisory control.

the physical system (note the two inputs to the estimation box) to get a best estimate of system state (or, more broadly, to "assess the situation"); and (c) detecting/diagnosing any abnormality.

4. If the operation comes to a point of planned updating, the supervisor *intervenes* to reprogram (solid horizontal arrow to the right). If there is an abnormality, the intervention is unplanned and represented in a different form (dashed line). Note that teaching and intervening essentially require the same subfunctions but are performed under different circumstances.

5. *Learning* is affected by (a) state estimation, (b) failure detection/diagnosis, and (c) command boxes — because learning is a matter of evaluating what happened in response to what the automation was asked to do.

The heavy arrows in the diagram represent the on-line supervisory activity, whereas the thinner arrows represent the off-line activities.

Various researchers have proposed different interpretations and augmentations to the supervisory control model. Murray and Liu (1997a, 1997b), concerned that in many supervisory control situations some inter-actions between elements are ill-structured, proposed to model for a network of conversations for dealing with exceptions in a background of otherwise routine behavior. Kirlik (1993) discussed a supervisory control experiment from the viewpoint of Heidegger (1977) and Gibson (1979), whose writings formed the background for the ecological perspective (discussed in the section on ecological displays in Chapter 5).

When to Automate (and Use Supervisory Control)

In general, the more complex a task is, the more time it takes to do. Assuming some measure of complexity, the time required for direct manual control might be represented by the dashed line in Figure 6.6. The time to complete any given task under supervisory control requires the time spent planning and teaching the task to be done (the thin curve in Figure 6.6) plus the time to execute the task (the slice between the thin curve and the heavy solid curve), so that the sum of these two times is represented by the heavy solid curve.

Note that the heavy solid curve intersects the dashed line in two places. This is because the dashed line begins at the origin; tiny, trivial tasks are performed manually in tiny, trivial amounts of time. That is not true of supervisory control, because teaching the computer even the smallest task requires a minimum amount of time overhead, even though computer execution may take essentially no time. It is much faster to do a simple task by yourself than to explain that task to a computer (or to another intelligent agent or person). It is faster to scribble a short note by hand than to open up a word-processing program and type it. Supervisory control is

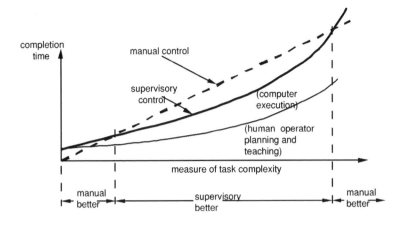

Figure 6.6 Range in which supervisory control outperforms manual control.

faster than manual control only after the task complexity has reached the point that computer execution speed more than compensates for the planning and teaching time overhead.

The right end of the scale is a different story. Very complex tasks are hard to figure out how to program. This is seen when trying to program robots for complex teleoperation tasks. At some point it really is quicker for those tasks to be done manually, which is why we pay the cost for astronauts to repair space telescopes in extravehicular activities. No one has any idea how to program computers to take care of children, write symphonies, or manage corporations.

If a task is to be repeated many times once it is programmed, and the environmental conditions do not change, the planning and teaching time can be amortized over the whole series of repetitions. This is true in manufacturing, for example, where for a large-batch operation, automation (and supervisory control) is economical, but for very small batches or one-of-a-kind fabrications, the work is best done by hand.

Chapter 7

SIMULATION, TESTING, AND EVALUATION

Part II, "Design of Human-Automation Systems," began with a look at the process of analysis and design of these systems (Chapter 3). Task analysis and function allocation were discussed in detail. Chapter 4 then examined various aspects of human performance as related to such systems, Chapter 5 considered various aspects of designing displays and decision aids, and Chapter 6 dealt with supervisory control. This last chapter of Part II is about the final stages of the design process: simulation, testing, and evaluation

Human-Automation Design as an Empirical Activity

Whoever speaks of optimizing a human-machine system is talking nonsense. At least in the formal sense of the term, *optimization* requires a mathematical equation representation of the system, the values of which can be adjusted to maximize a given global objective function. Unfortunately, neither the equation representation of the system nor the global objective function is available for any real human-automation system. The major reason for this is the presence of the human. Physical machines can be optimized to a reasonable degree with respect to well-defined criteria, simply because some well-behaved deterministic laws of physics are available. That is simply not true for human beings. Progress in neurobiology is exciting, but no one can say when science will come close to understanding and predicting living systems at the level that machines can be understood and predicted.

So, as in the field of medicine, the analysis and design of human-machine systems is empirical, and one must be satisfied with an experimental approach, making what generalizations one can, with numbers when possible, and with words otherwise.

Theory has its place, especially to provide guidance for experimentation. I believe that normative information, decision, and control theories are particularly salient, as they do lend themselves directly to experiment with human participants interacting with machines; this is the reason for their inclusion in Appendix A.

Sometimes experiments are done in the laboratory, and sometimes they are field tests. The laboratory offers the advantage of experimental control, but its disadvantage is that it is not the real world, and many of the real variables are not present. The researcher is advised to work at (and with) both laboratory and field tests if possible (not necessarily in that order), as there are different lessons to be learned from each. Ultimately, of course, the real world, the real marketplace, the real traffic, or the real battle is the arbiter of success. One problem of experimenting in the real world is the time delay before one gets useful feedback.

Human-in-the-Loop Simulation

Perhaps the most important way to experiment with human-automation systems is through simulation. Computer simulations of completely physical systems are used commonly for all of engineering and science, but here I am referring to simulators of the physical system — including displays, controls, automation, and environment — operated by human participants. Such simulators are called *human-in-the- loop simulators.*

Human-in-the-loop simulation technology has been used for many years in aviation, space, nuclear power, and the military for training in concepts, procedures, and skills as well as research and engineering development. Equipment failures and accident situations of all kinds can be simulated, and the human response can be tested. Such simulation is a common method to test operator and system response under circumstances deemed too unsafe to test in the real world (in flight, in space, in actual nuclear plant operations, or in battle). These simulators are also used for verification and quality assurance for new systems, for training, and for marketing.

Such use of simulators is less extensive in rail, marine, and highway transportation; medicine; and other fields requiring operator skill, but that picture is now changing. With computers becoming simultaneously cheaper and smaller, and the public becoming less tolerant of risk taking with experimental test participants, human-in-the-loop simulation is becoming increasingly accepted and economical.

Different Levels of Simulator Fidelity

At the low end in fidelity are desktop computer trainers with simple graphics. These are usually used for teaching concepts and are augmentations of books or lectures. Sometimes many simulator stations are used simultaneously in conjunction with a lecture. Computer graphics, when used at all, can be relatively abstract.

One step up are *part-task simulators*. These have some moderate level of fidelity for particular subsystems or particular operations to be exercised (i.e., when the user operates the correct control, the appropriate system response occurs). The other controls and displays, when it is useful to have a background context to the part task of interest, are typically mocked up with drawings or decals. Often these nonessential controls and displays are simply omitted. The part-task simulator is limited in size by not including large portions of the control room, flight deck, or workstation. Part-task simulators, once mostly electromechanical, are increasingly implemented with computer-based virtual displays and controls.

A further step up is the so-called more or less full-task simulator with only moderate fidelity. It is "more or less" because a few subsystems may be missing, and moderate fidelity means that it is dynamic and realistic but that time and pixel resolution may be limited and motion cues may be missing. Many general aviation simulators and driving simulators fit this category. They have all the essential controls, and you can fly the airplane or drive the car, look out the window, and get some feel for the dynamics, but you know it is not real.

The criterion of the true high-fidelity simulator is that it is so realistic that the participant easily forgets he or she is in a simulator. Accurate motion cues are there. The response dynamics are imperceptibly different from the real thing. The pixel resolution and field of view are such that the participant is truly immersed. The price tag is also high, mostly because of the motion base (in vehicle simulators), though immersive display graphics have become much cheaper in recent years compared with the electrohydraulic systems required to move the vehicle with the pilot or driver inside. For a process or manufacturing plant simulator, no motion cues are necessary (at least for normal operations!). The process plant's training control room simulator is typically a carbon copy of the real one, except that it is connected to a computer instead of the real plant.

How Much Fidelity Is Needed?

Everyone is impressed with high-fidelity simulators. They are fun for the experimental participants. Their owners are proud to show them off to managers, politicians, generals and admirals, other important visitors, and family. They help bring in money because they imply that serious research or training is being done. They are also expensive and require considerable maintenance, but that somehow gets figured into the budget; they are just part of building the simulation or training empire.

How useful are high-fidelity simulators? That is a different question, and the answer depends on many things. For some legitimate purposes, nothing but high fidelity will do. One example is final integrated training and testing exercises for commercial or military pilots in new (to them) aircraft or major new subsystems. Usually the first time a pilot who has undergone such training flies an actual new aircraft or new system, he or she has a full load of passengers. Another example is research on human driver decision/response in complex highway accident situations in which the subtle motion or visual cues in question cannot be simulated on the low-fidelity simulator. A third example is use of the high-fidelity simulator to calibrate a low-fidelity simulator — for example, to determine whether the lower fidelity is sufficient or yields results different from those of the high-fidelity simulator for particular performance measurements.

For most training or research, low fidelity will do just fine. Certainly for concept training, simple and abstract representations of objects or events are sufficient. Part-task simulators are the proper means to teach procedures having to do with a particular subsystem.

Major training establishments, such as commercial aircraft companies and military services, use a variety of simulators ranging from very simple to very high fidelity. Usually training progresses from the simple and low fidelity to the integrated, complex high fidelity.

Do Human Participants Act the Same in Simulators as in the Real World?

This question is often asked by people who oppose simulator research for one reason or another. The answer is "yes" in the high-fidelity simulator because operators associate all the cues with reality and let themselves be immersed. They feel the stress and perspire. Only when the simulation exercise is ended do they revert to feeling it is not real. If they were asked in situ during a training exercise, they would know the truth; no one is deceived.

For simple low-fidelity simulations, the question is not relevant if the simulator is being used for early stages of training. For research there may well be differences in behavior from that in the real world, but there is usually no alternative if controlled experiments are to be performed that would not be allowed in the real world or if the research budget is limited.

A High-End Simulator: The National Advanced Driving Simulator

At this writing the U.S. Department of Transportation National Advanced Driving Simulator (NADS) in Iowa City, Iowa, is probably the highest-fidelity human-in-the-loop simulator in the world. As shown in Figure 7.1, it consists of a modifiable vehicle in which the test participant sits, surrounded by a projection screen. All this is mounted through high-frequency vibrators on a 330° yaw ring, which in turn is mounted on a six-

Figure 7.1 Artist's rendering of the National Advanced Driving Simulator. Cutaway view of device in motion bay showing passenger sedan cab during a passing maneuver. (Photo courtesy L. D. Chen and the University of Iowa. Copyright © 2000 The University of Iowa.)

degree-of-freedom hexapod or Stewart platform (common in moving-base flight simulators). The platform is mounted on a large *x, y* plotter allowing 64 feet (19.5 m) of translation in both directions. In all, it provides displacement, velocity, and acceleration over a wide range, sufficient to simulate almost any highway vehicle maneuver or accident.

A high-resolution (5-million pixel) 360° visual scene is projected on the screen (15000 polygons refreshed at 60 Hz) under program control. A correlated high-fidelity surround-sound audio system is also provided, as well as control feel at 60 Hz. Sophisticated software and computer architecture allow realistic accident scenes to be generated and coordinated with the vehicle motion, sound, and control feel.

This simulator is now being used to investigate a variety of questions about how aging, physical handicaps, highway conditions, other driver behavior, drug use, use of car phones, use of navigation systems, or use of other auxiliary devices while driving affect safety. It also assists in the engineering of new safety aids such as intelligent cruise control, radar collision-warning alarms, and so forth.

Estimation Theory as a Metaphor for Using Simulation in the Research and Design Process

Figure 7.2 diagrams the process of estimation, as commonly considered by control engineers and estimation theorists. The boxes represent operations that in various applications are well defined and usually mathematical, typically linear differential equation transfer functions. However, in this case I have made changes to the usual labels, though the functions remain the same.

To the left of the dashed line are the functions that comprise what is usually called the *estimator;* in the present metaphor it is the research/ design process. The blob to the right is the real human-machine system (or the prospect of it if the system being researched/designed is a new one). Before the fact, a full understanding of the true reality (how any design will fare when put into the real environment) can never be known; it can only be estimated because of filters that limit what can be measured and what actions can be taken to test or, eventually, to influence the real world with the new system design. The arrows represent causality or signal flow, and their letter labels represent the variables (actually, vectors of many variables).

At the upper right, x is the state vector of interest, that subset of data characterizing the reality one seeks to understand and design for. It is measured by an imperfect measuring process, producing y, and the act of measurement has some effect x' back on the reality. For the most part, x'

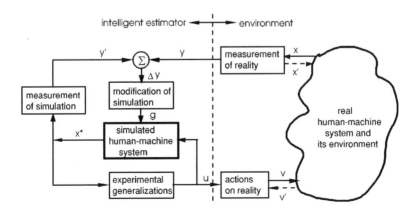

Figure 7.2 The process of estimation applied to human-machine system research and design.

is a small effect, but one can never measure something without affecting it: That is the Heisenberg uncertainty principle. In social interactions, x' is the effect the observer has on the observed person or organization.

On the left side, x^* is the set of data emerging from a simulation, in heavy outline. (I will say in a moment how the simulation comes to be.) This x^* is measured by means parallel to those that would be used in measuring x from the real world, producing y'. Insofar as there is a difference, Δy, between y and y', that is an indication that the simulation needs improvement, and so it is modified on the basis of the discrepancy with changes g to the simulation.

What drove the simulation to produce x^*? The answer is that knowledge of recent x^* suggests actions u to be taken to implement modifications v so as to achieve improved performance in the real world, and this same action is used as an input to the simulation. (At early stages, actions on reality may be inhibited by reactions v'.)

Gradually, by this incremental improvement procedure one improves the simulation model of the real-world interactions (those of interest) until the Δy is small. This estimation scheme forms the basis of modern control theory and is sometimes called an *observer* or *Kalman estimator* (after Kalman, 1960). Although here the term is used as a metaphor for research and design, others have used it as a metaphor or model of human behavior.

This estimation metaphor suggests that design is an iterative improvement process, a suggestion with which any designer would agree. One starts with a poor model or simulation of the system functioning in its environment and tests how design parameter changes will improve its performance. (In real design practice this is done first on paper and later as computer simulation.) After multiple iterations of changing the design and exploring how it might succeed in the real world, as best as one can test that out, one eventually converges on a final design.

Caveats on Designing Experimental Tests

Experimental design is a whole discipline by itself. At the risk of making experimental design seem oversimple, I mention here a few terse caveats about experimenting with human participants; this is territory that the psychologist reader is surely familiar with, whereas the engineering reader may not be. Readers who do not understand these reminders are advised to consult a reference on the design of experiments, such as Cook and Campbell (1978) or Kappel (1992). This is important in deciding on the particular experimental treatments or conditions to be set, the selection and number of participants, the method of training and instructions to participants, the statistical tests to be used, and the conclusions that can be drawn, including the validity and generalizability of those conclusions.

Posing the Experimental Question

The fact that a human-automation system is complex does not excuse the researcher from devoting some effort to posing the experimental question in simple and direct terms: What performance (dependent) variables are of interest, and what independent variables (both controllable and uncontrollable) are of interest to test for their effects on those dependent variables?

The Importance of Preliminary Experiments

Preliminary experiments, often with the researcher or a convenient colleague as participant, are always useful prior to committing to gathering statistics in a formal experiment with multiple participants. By the criteria of time and scientific discovery, there is much to be gained by thoughtful preliminary testing to refine the experimental question, What variables and what ranges of those variables make a difference? When the formal experiment is begun, one should have a rather good expectation of the results, which presumably the experimental results will verify.

Experimental Design Logistics

There is a tendency to launch into an experimental investigation by designing factorial experiments with too many dimensions (independent variables, or *treatments*) and too many levels of those independent variables. The researcher soon discovers that the number of combinations is huge and that time will not reasonably allow such an experiment to be done. Human-automation design is not epidemiology, and usually one is not looking to show statistical significance from small influences of an independent variable on a dependent one. One is looking for hardware, software, and procedural factors that make a large difference.

Participant Selection and Training

The power of a statistical test — the probability of rejecting a null hypothesis when it is in fact false (see Appendix A) — improves with the number of participants, so all other things being equal, having more participants is always better. However, given constraints on time and money, having a smaller number of participants who are well trained in the experimental task (but who are given no hint about what the experimenter expects the results to be) is often sufficient and, in any case, is better than having many untrained participants. So one must consider these trade-offs (detailed in references such as Cook and Campbell, 1979, and Kappel, 1992) at the outset of experimental design.

It is important that the participants in the formal experiment represent the population to which generalizations need to be made. I once used secretaries and company engineers as participants in predicting how well *Apollo* astronauts would perform a required visual-motor skill task and found later, in testing the astronauts themselves, that they were far better at the task than could be predicted from the unsophisticated participants.

Controlling Effects of Learning

One should counterbalance the orders in which different participants are exposed to different treatments (e.g., Participant 1 is assigned the order *ABC*, Participant 2 is given *BCA*, and Participant 3 gets *CAB*) so that learning effects, which tend to occur in spite of training, tend to cancel out when the averaging is done.

Analysis of Results

There is a danger in using widely available statistical packages to "prove" statistical significance without some understanding of the assumptions inherent in those statistical tests. Also, a reminder is in order that 5% of the time, no systematic effect (randomness) will produce a statistically significant effect at the 95% confidence level. There is also a danger of overgeneralizing results. The results are valid for the particular conditions of the experiment performed. It is always safer to claim too little, to hedge the wide applicability or results, than to overclaim.

Virtual Engineering

The Virtual Reality Bandwagon

Computer-based *virtual reality* (an oxymoron; *virtual environment* is better but hasn't caught on) is not only in high fashion, it is also truly significant for the design of human-automation systems.

The activity surge of recent years has been brought on by dramatically increased computer capability: higher computational speeds, greater memory, and higher display resolution at a lower price. In addition are the new technical developments in head-mounted displays, which generate their display based on where the head is pointed; stereo goggles; and auditory and haptic counterparts. These promise increasingly immersive involvement of the user in system operation. Currently the primary economic driver is not research, training, remote control, or human-automation systems, but entertainment: computer games and movie animation.

In the engineering design world there is a whole new array of software packages to do computer-aided design, computer-aided manufacturing, and computer-aided control. The Boeing 777 was allegedly designed essentially by computer, and engineering students are taking for granted powerful analysis toolkits such as MatLab and ProEngineer.

Human-automation systems are becoming more amenable to design by virtual reality. Not only can the automation be simulated (that is usually the easy part) but, to a limited extent, so can the human body. Computer graphic male and female figures, representing both ends of the size range, can be programmed into computer graphic workplace renderings to test for visual, reach, and other ergonomic compatibility considerations. Visual workload and control dynamic compatibility models can also be inserted. Gradually these graphical humanoids will be able to act dynamically, allowing the evaluation of forces on the body. It is a modeling discipline very much in flux.

The Delusions of Virtuality

TV advertisers and moviemakers are using computer animation to achieve a large degree of virtual reality, and the same techniques are being used in courtrooms to convince juries of "what really happened"(when in fact the virtual reality can be just as scripted and biased as an oral argument). There is a similar danger that designers come to feel that because a designed system looks good in virtual reality and passes some tests of human-in-the-loop simulation, it is ready for service and/or the marketplace. I have observed this with my own students. This is a dangerous delusion, because usually many unseen and unexpected factors appear when field tests are carried out and a system comes into use in the real world.

Evaluation

What Is to Be Evaluated, and by Whom

The most human task in both designing and operating a human-automation system is evaluation. Table 7.1 summarizes what is to be evaluated by the designer and by the operator.

Both the system designer and the system operator must have firmly in mind the mission objectives and interpret these in light of the circumstances that face them. Neither the designer nor the operator has free rein in making evaluations during design or operation, respectively. Most likely, certain standards have already been set by the regulations, management policies, laws, and norms of society as to what behaviors are expected and what are efficient and safe operating practices. The designer

TABLE 7.1
Evaluation Responsibilities of the System Designer and the System Operator

Responsibility of the Designer	Responsibility of the Operator
Evaluation of mission objectives	Evaluation of mission objectives
Evaluation of operating standards, regulations, laws	Evaluation of operating standards, regulations, laws
Evaluation of capability of prospective operators	Evaluation of variables in real-time system operation
Evaluation of objective function for automation	
Evaluation of training objectives	

must have these constraints in mind when deciding what the task objectives are (task analysis) and what the function allocations are between humans and machines.

The system designer has the task of prospectively evaluating the capability of the prospective human operator to perform the tasks required of him or her by the system design. The operator who presumably meets the designer's criteria does real-time evaluation during operation. This includes evaluation of plans, of what specifics to teach the computer to implement those plans, of how the automation performs, of any failures and what to do about them, and of his or her own experience. These are not tasks the computer is good for, except as a record keeper and an analyzer.

The designer also has the formidable task of specifying the objective function of the automation. Within its own constraints on speed, accuracy, and energy efficiency, automation can be designed to control to whatever objective function is specified. In older forms of automation there was little flexibility in automation; feedback controllers merely nulled error. Modern computer-based automation permits much greater selectivity. The designer can make up the rules, have the set points and criteria be adaptive to the environmental circumstances, and so on.

Because the operator must undergo training, the designer also, at least implicitly, sets the training objectives. (A training expert will later interpret these objectives to set a specific training regimen.)

Techniques for Evaluation
Many techniques have been employed to accomplish evaluation of designs and policies. Some are for individual judges' evaluations, others for evaluations produced by groups. Appendix A summarizes a recognized formal technique for a single judge to determine what design alternatives are not dominated by other alternatives in terms of salient criteria or objectives, where selection among the remaining alternatives is then done by elicitation of single- or multiple-objective utility (relative worth) functions.

The following paragraphs contain several less formal techniques by which groups of judges can evaluate design alternatives.

Use of *focus groups* is one popular technique. This is essentially a free-form discussion, often around a set of questions posed by the group convener. Participants are drawn from various demographic or socioeconomic sectors of the user community. The problem with a focus group is that it can be dominated by one or a few people, not necessarily those with the best insights.

Group voting (rank orderings, qualitative category ratings, etc.) is useful provided that the questions to be voted on are well stated, but evidence from polls shows repeatedly that wording a question in a slightly different manner produces a different voting result. A means around this is to use electronic anonymous voting (whereby the group sees an instantaneous electronic tally of how many, but not who, voted how), thus relieving any reluctance to participate. If this is followed by open discussion to repeatedly modify the question and revote, the group quickly comes to understand what it is that changes the vote and what is most important to the participants (Sheridan, 1975).

In another technique, known as the *Delphi method,* the judges make their numerical or rank-order evaluations, state some reasons why, but remain anonymous as results are shared among the group (Dalkey & Helmer, 1963). On subsequent rounds participants question one another's premises and are allowed to change their votes. Only after several such rounds do they come face to face and try to hammer out a consensus.

Usability testing is a term applied to a range of techniques to evaluate user performance and acceptance of products and systems, which can be as diverse as Internet Web pages and application software, medical devices, and amusement park rides. In some instances usability testing collects behavioral data in a goal-oriented task within an applied scenario and in a naturalistic environment (home, office, etc.). In other instances focus groups, competitions, and other means are employed that do not use experimental controls. Usually the number of participants is small and the studies are relatively brief, as compared with most laboratory research. Wichansky (2000) reviewed the current status of usability testing.

A final method to be mentioned, a more quantitative one, is called *multidimensional scaling.* In this instance the judges do not give overall quality scores or preference orderings but, rather, specify numerically the perceived "distance" or "difference" between all pairs of alternative objects or events under consideration, considering anything that comes to mind that makes them different. A least-squares calculation (Shepard, Romney, & Nerlove, 1972) results in a plot of all the alternatives in a several-dimensional space (usually a much smaller number of dimensions than the number of judges). The dimensions of this plot are the abstract factors that produced the perceived differences (accounted for the greatest variability). The challenge then is to infer the meaning of the major dimensions or factors.

The problem of determining group preference, or *social choice,* has an interesting history in economics. Many theoreticians have pondered what is the best way for a group to rank order a set of alternatives. Arrow (1963) won a Nobel prize for, among other contributions, proving that there is no procedure by which more than two people can achieve a consensual rank ordering of alternatives without some of the time being intransitive — what has come to be called the *impossibility theorem.* (A trivial example of intransitivity is when Judge 1 orders the alternatives *ABC*, Judge 2 orders them *BCA*, and Judge 3 orders them *CAB*. Thus by majority vote there is no winner.) Arrow's more general proof was based on a set of now famous and seemingly obvious assumptions: (a) that the social ordering is based on the individual orderings; (b) that if all judges prefer *A* to *B*, the society prefers *A* to *B*; (c) that judges cannot defer to one who acts as dictator; and (d) that preference between any two alternatives is independent of feelings about some other alternative. However, given Arrow's theorem (which fortunately seldom matters in "majority rules" voting for a single winner with enough voters), democracies plod on, and groups of people continue to evaluate designs.

Conceivably, evaluation techniques such as those just mentioned could be used for on-line multiple-supervisor control of complex systems, assuming the dynamics of the process were slow enough to allow the necessary communication and deliberation between the human supervisors. For example, in command and control systems the evaluation of risks and other costs in planning or intervening, of alternative diagnoses of failures, or of lessons learned come to mind.

PART III: GENERIC RESEARCH ISSUES

Chapter 8

TECHNICAL ISSUES OF HUMANS AND AUTOMATION

In Chapters 8, 9, and 10, I discuss problems of humans and automation from a broad and generic perspective, and for that reason the term *issues* is used. A wise colleague once advised me that the challenges in life were of two kinds: problems and issues. Problems, he asserted, have solutions; issues do not. One tries to solve the problems, and usually one does. With no less obligation one tries to make progress in dealing with the issues, succeeds a little, but knows that ultimately the issues will persist, though changing in form.

System Complexity: How to Cope

Distributed Systems and Hidden Complexity

Readers with only cursory knowledge of automation might get the impression that automation systems have one sensor, one decision element (computer), and one actuator tied together in closed-loop fashion and performing the control of one variable. That may seem to have been implied by the foregoing text, and indeed many automatic systems in the past were exactly that. It seems today that a safer assumption would be that humans typically have multiple automation loops to control (again, each loop with one sensor, one decision element, and one actuator system), each of which is independent of the others and most likely occupies its own unique space. However, as technology is developing, these notions of separateness and independence are misleading.

The truth is that automation systems tend to be increasingly distributed in space and mutually interactive. Figure 8.1 illustrates what is emerging. A system may have multiple sensors, multiple decision elements, and multiple actuators that are not necessarily contiguous, given that signal communication easily spans distance. Communication can be instant and high bandwidth or can be of low bandwidth with long delays. Several sensors may affect the same decision element. A decision element may drive several actuators.

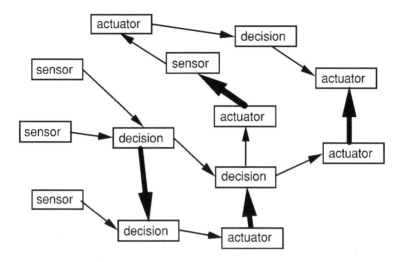

Figure 8.1 Coupling among sensor, actuator, and decision elements within a system. Blocks can represent machine or human elements. Heavy lines suggest unintended cross-coupling.

Most of these are probably intended, engineered interactions, but there may also be unintended interactions. For example, one decision element may subtly affect another decision element (e.g., one person's decision is influenced by another person's decision). In addition, some actuators, through coupling and feedback, may unintentionally affect sensors, decision elements, or other actuators. In Figure 8.1 these last interactions are shown by heavy lines. This is a vignette of what systems engineers and physicists are calling *complexity.* Anticipating and understanding all these interactions can become almost impossible.

Typically, mathematical models are not formulated beyond some modest level of complexity. For example, the full nonlinear dynamics of a six-degree-of-freedom robot arm cannot be modeled as a single dynamic system in all degrees of freedom. It is simply too complex, even though this may seem like a relatively simple mechanical system. When it comes to a manufacturing plant, a hospital, or a multiple-aircraft or multiple-car traffic interaction, comprehensive dynamic modeling is simply beyond the pale.

Add to this picture of complexity the fact that large-scale systems are constantly changing, so that a diagram such as Figure 8.1 is different from one month to the next. Rasmussen (2000) characterized such large-scale human-machine systems as being dynamic in terms of (a) pace of change, (b) scale of operations, (c) integration of operations, (d) aggressive competitions between elements, and (e) deregulation by government.

Fixes to Fixes

When accidents or other unwanted or unwonted events occur and people are hurt, bodily or economically, well-intentioned operators and designers try to fix things. Driven by the pressures of the marketplace, the media, and particularly government regulators, they are motivated to do it quickly and without large additional expenditure. This creates a problem, that of superficial fixes.

For example, in recent decades in nuclear power plants, when malfunctions with the automation occurred that threatened safety, there was a tendency to try to improve safety by adding more automation. This might have corrected the original problem, but it often produced other problems, not the least of which was greater complexity and less predictability and understanding by the human operators. In this regard it has been pointed out that Three Mile Island Unit 2, which suffered its core meltdown because operators were confused about what was happening, was significantly more complex a plant than the older Unit 1 next door, which did not have so much automation and updating. (With trepidation, I participated in some postaccident human factors fixes to Unit 1!)

Mixed Human and Computer Initiatives

As automation systems become more complex, with more decision elements (as noted in the last section), there will be more controller set points, more rules, and more embodiments of the objective function for automatic control — not all colocated. This is what has come to be called the *mixed-initiative* problem: Different parts of the system try to pursue different goals, and as a result the automatic actions conflict with one another. Formal recognition of the mixed-initiative problem in automation is relatively new.

Most large-scale human-automation systems are operated or supervised by teams of people. Examples are many: the flight crew of an airplane, air traffic controllers, the crew of a ship, the conductor and engineer of a train, the surgical team in a hospital operating room, the control room operators in a nuclear power station or chemical plant, a construction crew, and various teams in police, fire, or military operations. Each team member has his or her own idea of what the objectives of the mission are, which results in conflicting actions. Of course, interpersonal disagreements arise in any organization of humans, to which highway traffic snarls attest (in this case the vehicles are benign).

When mixed human initiatives combine with mixed automation initiatives within the same system, however, this presents a particularly challenging problem. There is not much experience as yet for engineering this problem, but with increased complexity in human-automation systems, there is bound to be much more of it. What may be called for is a human-machine overseer (a metasupervisor?) who/which closely monitors the system for any conflicts and, when they are spotted, goes into action to inhibit some control loops or human wills and lets others continue unabated. In artificial intelligence this is called a *subsumption architecture;* the low-level control loops are subsumed under the meta-agent control.

Monitoring: A Job Nobody Wants

Bandwidth Limitations

I have noted several times earlier in this book that people are poor monitors of automation. If the automation is working well, people become bored and inattentive and fall sleep. People's physiological needs dictate that they cannot monitor continuously in any case. When failures or unexpected circumstances occur in complex systems, there can be sudden information overload and ensuing chaos while the human tries to cope. The human nervous system is somehow relegated to a range of bandwidths that is much narrower than the range in which automation can be put to work. At the low frequencies humans fail statistically, and at high frequencies, above 1 Hz, they fail reliably. Nevertheless, by default, human operators are assigned the task of monitoring.

Complacency?

Human-computer cooperation in monitoring would seem to offer much promise, particularly if process models are available (see the earlier discussion with respect to model-based prediction and model-based failure detection) and if ways are found for operators to buy time for their own slow nervous systems. Commercial aircraft pilots, for example, find it necessary to organize their tasks so that different tasks do not demand attention at the same time and the pilots can self-pace at a comfortable rate.

Can systems be designed so that no monitoring is required? Only with sufficient reliability over the range of circumstances anticipated by the design. Will this lead to operator complacency? That is probably the wrong term for an acceptably reliable system. Under the happy assumption of sufficient reliability, operator inattention *is* the best strategy. People do not waste time monitoring whether the ceiling will cave in over their heads. One need not attempt to maintain situation awareness of those variables that seem irrelevant.

Robustness is desired in automation, and human operators are included to augment robustness. In systems in which the public is at risk, it is asserted that they should "expect the unexpected" — but at what price? There are limits to operator expectation and awareness.

Giving and Taking Advice

When Decision Aids Are Least Feasible

In unfamiliar or unexpected situations, a fast response is essential. Experienced operators are forced to employ whatever automatic or "recognition-primed" behavior can be brought to bear. Inexperienced operators may panic. It might seem that decision aiding is an obvious remedy for a novice in a situation in which response is not automatic. Upon a little reflection, though, one realizes that decision aids and automation are least reliable when situations cannot be predicted and modeled by the designer, and the common solution to this is to keep a human in the loop.

Smith, McCoy, and Layton (1997) called this shortcoming of decision aids *brittleness* and showed in an air traffic control context how brittleness of an aid can influence a human to make poor decisions. Further, the burdens of using decision aids under time stress can outweigh the benefits, even if they work (Kirlik, 1993). The designer is left in a catch-22 ("a problematic situation for which the only solution is denied by a circumstance inherent in the problem," *Webster's 10th Collegiate Dictionary*): to use a decision aid or not? The answer surely depends on the mix of operator experience, decision-aid sophistication, and complexity of momentary circumstances.

Undertrust and Overtrust

It is easy to argue that control of air and highway vehicles and other safety-critical systems should never be given over fully to an automatic system. The human operator should always be there — at least to override the automation in case of failure or unexpected circumstances. If an alarm or warning system gives too many false alarms or a decision aid gives advice that proves incorrect or irrelevant, it is only natural for the operator to decrease his or her trust, eventually perhaps ignoring the alarm, warning, or decision aid (Lee & Moray, 1994). There is a tendency, usually rational, to be somewhat distrustful of a new warning, alarm, or decision-aiding system until it has proven itself.

To the extent that events are sufficiently measurable, predictable, and modelable, then the design of trustworthy warning, alarm, and decision-aiding systems is possible. Given a trustworthy system, a well-trained operator, and sufficient time for the human to receive the signals or advice, one might assert that decision aids should always be used. Then it is a small step to recommend that the operator be given whatever advice he or she wants. The operator can always reject the advice.

However, this philosophy is naive. Repeated use of reliable decision aids unavoidably leads to trust in the decision aid. Trust leads to dependence. Dependence can lead to slavish obedience to what the decision aid recommends. At some point of conditioning, the operator in effect abandons responsibility for his or her actions — though not formally or in theory. One can easily delude oneself into thinking the operator, even though well trained and alert, will always act responsibly.

It is only natural for a human to be intimidated and back off when encountering an apparently superior intelligence. Therefore the design engineer and the operations manager must think carefully about how much, and when, a decision aid should give advice and how to instruct the human operator about how to regard such advice.

The better the operator's understanding of how the automation and/or decision aid really works in different scenarios, the better able he or she is to maintain the proper level of trust. There is always a compromise between enough trust to make effective use of the advice and enough distrust to stay alert and know when to ignore the advice.

The Roseborough Dilemma

A theoretical problem in the engineering of decision aids has consistently frustrated those who seek an objective basis for designing the human operator into complex technological systems. One comes upon the dilemma as follows:

1. In a complex control system, the controlled process cannot be fully and explicitly modeled, nor can the objective function (as has been duly noted earlier).
2. Therefore, as has been asserted in previous chapters, one appropriately falls back on the human operator to complement whatever mechanized embodiment of these functions is provided.
3. Given that the human operator makes mistakes, it is evident that he or she can be helped by a computer-based decision aid, so one tries to provide such an aid.
4. Ideally, to design the decision aid and evaluate the human operator's use of it, a relatively complete process model and objective function must be used as a norm. However, Step 4 is in conflict with Step 1.
5. Further, if such a relatively complete process model and objective function were available, then why not use these in place of the human operator to provide an automatic decision maker, thus leaving out the human?

Roseborough (1988) concluded that "in any system requiring a human operator, the objective validity of a specific decision aid can never be established." On the basis of the line of reasoning just given, this statement is self-evident *if* the decision aid is intended to be used for decisions requiring information that is not explicitly modelable. However, a way out of the dilemma is to assume the following:

1. The human decision maker is necessary for the information that is not explicitly modelable. No valid decision aid can be built to provide such information when it is needed.
2. In some, perhaps most, decision situations the human operator will encounter, he or she will require only information that is modelable. The human will always make some mistakes in such decisions and can benefit from a decision aid for these cases, and in such cases the decision aid can be validated. (So it makes sense to provide such a decision aid for those situations.)
3. The human can properly decide when the situation includes elements the decision aid can properly assess and can know for which elements the decision aid should be ignored.

Assumption 3 is a big one, and it is an issue in obvious need of research.

How Far to Go with Automation

The All-or-None Fallacy

The nontechnical public tends to see automation as all or none: A system is controlled either manually or automatically, with nothing in between. For example, in the case of robotized factories, the media tend to focus on the robots, giving little mention of various design, installation, programming, monitoring, fault detection and diagnosis, maintenance, and learning functions that are performed by humans. In the space program the same is true: Options are seen to be either "automated" or "astronaut" without much appreciation for the potential of supervisory control. The truth, as has been discussed extensively in the preceding sections, is that humans and automatic machinery are, and will continue to be, working together.

Figure 8.2 helps in considering the state of progress relative to various degrees of automation and to the complexity or unpredictability of task situations to be dealt with. The meanings of the four extremes of this rectangle are worth considering.

The lower left is labeled "drudgery, slavery" because to employ a human being to perform completely predictable tasks is demeaning (although the truth is that many of us voluntarily operate pretty close to this when doing many small tasks each day). The upper right, "ideally intelligent automation," is not attainable in the foreseeable future, and one can argue that this is just as well because robots that clever would be a threat to humans. The upper left is where most people feel humans belong — working on undefined and unpredictable problems. Indeed, this seems to be where creativity and dignity, at least of the intellectual sort, are to be found. The lower right, in contrast, seems an entirely appropriate locus for

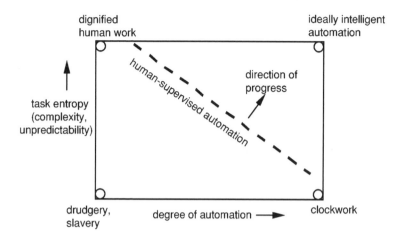

Figure 8.2 The trend of progress in human-supervised automation.

full automation; however, none but the simplest fully automatic machines already exist.

Few real situations occur at these extremes. Human-supervised automation may be considered to be a range or trade-off of options (diagonal line) advancing gradually toward the upper-right corner with technological progress. This is the Pareto frontier (see Appendix A) for human-automation systems.

Where to Stop Automating

Chapter 3 presented a scale of degrees of automation. It gave no simple answers regarding how far to go along that scale toward full automation with no human involvement. There are no simple answers. The answer obviously depends on many criteria.

There is some credence to the notion that automation, if reliable, relieves the human and can prove to be safer and cheaper. However, each more-automated level carries with it additional opportunities for machine error and precludes human intervention to a greater extent. To what extent should the responsible designer accede to the imperative of technological progress — to automate whatever can be automated and let the human handle what remains? Or should automation, at some level, be curtailed on principle?

Clearly, for some tasks, designers are happy to let the computer go to the last level, whereas for others they would prefer to limit automation to little or none. We, as designers, may give as a reason that systems are more safely operated with people in charge (at whatever level). What we may mean is that we trust people more because of a lack of understanding or inability to predict what the machine will do. Scientifically, however, it seems that machines operating on the basis of known, stored programs should be more predictable than humans. Maybe it is because, as human designers, we have more empathy for humans than for machines and can predict the kinds of exogenous circumstances in which humans can at least *cope* — even though we cannot predict exactly how they will cope or exactly what they will do.

Alternatively, the real motivation for deferring to human operators may be sociopolitical — other people will be more accepting of that decision. Evidence of these motivations can be seen in both the nuclear power and the commercial aviation industries, where acceptance of computers has been limited even though the technology might be capable of far more. Meanwhile, there are serious efforts to determine numerical reliability for various alternatives, ranging from fully manual to fully automatic.

The tendency, for obvious reasons, has been to automate what is easiest and to leave the rest to the human. From one perspective this dignifies the human contribution; from another it may lead to a hodgepodge of partial automation, making the remaining human tasks less coherent and more complex than they would otherwise be, and resulting in overall degradation of system performance (Bainbridge, 1983).

Ultimate Authority: When and By Whom

Smart computer-control systems can blur the lines of authority. In some modern aircraft the pilot is barred from making certain critical attitude maneuvers because the designers have deemed that those maneuvers would put the aircraft into an unsafe condition. Pilots, however, complain that some circumstances call for such unusual maneuvers, so they assert that pilot authority should never be limited. Another example is when both pilots attempt to control a fly-by-wire aircraft by making very different responses on the controls and the current system, not knowing which is correct, is programmed to take an average. Obviously, in an avoidance maneuver, if one pilot commands a pitch up and the other commands a pitch down, an average results in a head-on crash. How to accommodate such factors?

The problem of authority between human and computer is one of the most difficult in new transportation systems. Popular mythology is that the human operator is (or should be) in charge at all times. In the past, when a human turned control over to an automatic system, it was mostly with the expectation that circumstances were not critical and that he or she could do something else for a while (as in the case of setting one's alarm clock and going to sleep). Even when continuing to monitor the automation, operators are loathe to "pull the plug" unless they receive signals clearly indicating that such action must be taken because the automation has clearly failed.

Authority Dilemmas in Aviation

Examples of some authority questions in the aviation field that have been or are now being debated are the following:

- Should there be certain states, or a certain envelope of conditions, for which the automation will simply seize control from the operator? In the Airbus, for example, it is impossible to exceed critical boundaries of speed, altitude, and attitude that would bring the aircraft into stall. In the MD-11, the pilot can approach the boundaries of the safe flight envelope only by exerting much more than normal force on the control stick.
- Should the computer automatically force deviation from a programmed control strategy if critical unanticipated circumstances arise? The MD-11 will deviate from its programmed plan if it detects wind shear.
- If the operator programs certain maneuvers ahead of time, should the aircraft automatically execute these at the designated time or location, or should the operator be called on to provide further concurrence or approval? The A-320 will not initiate a programmed descent unless that action is reconfirmed by the pilot at the required time.
- In the case of a subsystem abnormality, should the affected subsystem automatically be reconfigured, with after-the-fact display of what has failed and what has been done about it? Or should the automation wait to reconfigure until after the pilot has learned about the abnormality, perhaps been given some advice on the options, and had a chance to take the initiative? The MD-11 goes a long way in automatic fault remediation.

Going All the Way

If all the information required for full automation is available, why not automate fully? In the railway sector this question has been asked, and in some special cases the decision was to remove the driver entirely. As an example, there are now fully automated trains on rails or rubber-tired guided vehicles in some new airports and similar trains on dedicated tracks, such as that between Orly Airport and Paris. Many engineers assert that automatic control is essential for modern high-speed trains because speeds are too great for drivers to react to obstacles, and there is simply nothing more to debate. However, history has shown that with regard to automation, designers are not always as smart as they may think. Although automation is now widely accepted in aviation by pilots, airlines, and regulators, accidents have occurred that have been blamed on the automation design philosophy. In the case of the Airbus, there has been active discussion about whether automation has gone too far.

It is salient to note that I observed Charles Stark Draper, the so-called father of inertial guidance and of the *Apollo* navigation system that took the astronauts to the Moon, to proclaim at the outset of the *Apollo* program that the astronauts were to be passive passengers and that all the essential control activities were to be performed by automation. It turned out that he was wrong. Many routine sensing, pattern recognition, and control functions had to be performed by the astronauts, and in several instances they countermanded the automation and saved the mission.

A fatal accident occurred in 1996 on the highly automated Washington, D.C., Metrorail transit system. Experiencing many station overruns (the automatic speed controls were receiving unreliable data from track sensors because of ice and snow), a train operator repeatedly requested central controllers to turn off the automation and allow him to control the speed manually. He was refused, largely on the basis of a management decree that the automation was *not* to be turned off. At one stop he overshot again and crashed into a standing train at the end of the line. A 1997 National Transportation Safety Board report pointed to management as a prime cause (Parasuraman & Riley, 1997).

Potential Strategies

Assuming sufficiently accurate models of vehicles and their interaction with their respective environments, and assuming sufficiently accurate state measurements, optimal automatic control is feasible. Alternatively, as noted earlier, an optimization calculation can be made continuously and used, not for automatic control, but to display to human operators a best profile for human control action. In aircraft such a system is called a *flight director.* A comparable computer-based optimal trajectory advisory system for a train driver was described in Chapter 2.

Proponents of the latter kind of aid maintain that if the humans keep precisely to such a profile, they can in any case do a better job at control than trying, in effect, to perform complex calculations in their heads. In addition, the humans are there if some totally unpredictable events occur. This can obviate public fear of riding in an unstaffed vehicle and would also inhibit litigation, because if an accident occurred it could be claimed that both automation and human were present and doing their best.

As an alternative to a decision aid assisting the human, there is always the possibility of the human assisting the automation. Both strategies should be considered. Figure 8.3 is an example of how one might think about options for authority allocation as a function of how critical the situation is.

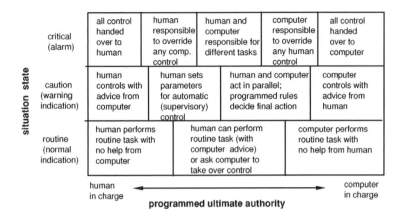

situation state		human in charge ← programmed ultimate authority → computer in charge

situation state	critical (alarm)	all control handed over to human	human responsible to override any comp. control	human and computer responsible for different tasks	computer responsible to override any human control	all control handed over to computer
	caution (warning indication)	human controls with advice from computer	human sets parameters for automatic (supervisory) control	human and computer act in parallel; programmed rules decide final action		computer controls with advice from human
	routine (normal indication)	human performs routine task with no help from computer	human can perform routine task (with computer advice) or ask computer to take over control			computer performs routine task with no help from human

human in charge ←——————————————→ computer in charge

programmed ultimate authority

Figure 8.3 Categories of authority allocation as a function of situation criticality.

Automation is sure to improve and to relieve the human of more tasks in the future. However, unsubstantiated claims — as are now sometimes made for how intelligent automobiles will increase safety and reduce congestion — must be guarded against. The potential is there, but empirical system demonstrations speak louder than promises.

Human-Centered Automation: What Does It Mean?

The term *human-centered* (or *user-centered*) *design* (or automation) has been widely popularized as the proper approach to integrating humans and machines in a wide variety of systems (Billings, 1997; Rouse & Hammer, 1991; Wickens, Mavor, et al., 1998). What does it mean?

Humans and machines differ in many ways, and yet they are similar in others. There is little disagreement with the idea that the ideal is to make the best use of both human and machine — in cooperation, harmony, integration, or whatever good word one chooses. This means the human interacting with the automation not only in system operation but also in design, manufacture, installation, and maintenance. A landmark paper by Licklider (1960) made the case for "human-computer *symbiosis,*" the latter word implying an essential interdependence of human and computer on each other.

With respect to differences between humans and machines, Chapter 3 cited the Fitts MABA-MABA list of those functions "men [sic] are better at" and "machines are better at." Although this list has motivated debate about how to think about and implement automation interactively with people, it has provided little in the way of concrete engineering guidance as to how to integrate people and automation in specific design instances.

Is Ecological the Answer?

In Chapter 5, I discussed the recent trend to see human response as being *recognition primed or naturalistic* (Zsambok & Klein, 1997), in much the way a neural net is triggered. This also accords with the tendency, as described by Rasmussen (1986), for learned behavior to move from being *knowledge based* to being *rule based* and, finally, to become *skill based.* Suchman (1987), writing about human behavior in interaction with machines, used the term *situated,* meaning situation or context dependent. All of this is sometimes placed under the rubric of what Gibson (1979) called *ecological,* referring to the fact that animals adapt their behaviors to accommodate to the signals and patterns they naturally find in their environments that directly affect their welfare.

In human-made technological environments, the idea of ecological design is to make it natural for humans to respond to patterns that are salient to the task so that these patterns can easily be perceived, understood, and acted on (Flach, 1998; Flach, Vicente, Tanabe, Monta, & Rasmussen, 1998; Vicente and Rasmussen, 1992; Vicente, 1999). A proposed ecological display for process control was mentioned in Chapter 5. *Human-centered* as meaning *ecological,* in this qualitative sense of serving the human in relation to the environment, is hard to dispute.

An Accepted Idea Searching for a More Specific Meaning

Human-centered design often conveys various, more specific meanings and different assumed design principles to different people. Some meanings, if taken literally and without qualification in designing automation or anything else, lead to trouble. For different meanings, one can find cases in which the assumed principle will not work and will lead to worse performance.

Table 8.1 lists 10 principles that have been stated in the past as meanings of *human-centered,* which at first glance seem like reasonable bases for system design. After each, the italics mention circumstances for which that principle is inappropriate (Sheridan, 1997).

TABLE 8.1
Nominal Criteria of Human-Centered Design (and Reasons to Question Them)

Allocate to the human the tasks best suited to the human, and allocate to the automation the tasks best suited to it. *(Unfortunately there is no consensus on how to do this!)*

Keep the human operator in the decision and control loop. *(This is good only for intermediate-bandwidth tasks. The human is too slow for high bandwidth and falls asleep if bandwidth is too low!)*

Maintain the human operator as the final authority over the automation. *(Humans are poor monitors, and in some decisions it is better not to trust them!)*

Make the human operator's job easier, more enjoyable, or more satisfying through friendly automation. *(Operator ease, joy, and satisfaction may be less important than system performance!)*

Empower or enhance the human operator to the greatest extent possible through automation. *(Encourage megalomanaic operators?)*

Support trust by the human operator. *(The human can come to overtrust the system!)*

Give the operator computer-based advice about everything he or she should want to know. *(The amount and complexity of information is likely to overwhelm the operator at exactly the worst time!)*

Engineer the automation to reduce human error and minimize response variability. *(A built-in margin for human error and experimentation helps the human learn and not become a robot!)*

Make the operator a supervisor of subordinate automatic control systems. *(Sometimes straight manual control is better than supervisory control!)*

Achieve the best combination of human and automatic control, where *best* is defined by explicit system objectives. *(Rarely does a mathematical objective function exist!)*

Limits of Modeling and Prediction in Human-Automation Systems

Modeling Cognition

One could say that usable models exist for some aspects of human sensory-motor skill — for example, for discrete and continuous target acquisition. It is harder to claim there are usable models of cognition, which is the principal function of humans supervising automation and is not sensory-motor skill. Unfortunately, the demand these days is for models of cognition, less so for models of sensory-motor skill. In aviation, for example, just as usable pilot models for manual aircraft handling were being refined, aircraft piloting shifted to pilot supervising of autopilots, but modeling pilot cognition within the supervising role has proved to be a much taller order than that of modeling manual skill.

The literature is full of cognitive science experiments, and they all have their validity in specific contexts. But the problem is that so many variables have an effect on what people perceive, remember, think, and decide that it is difficult to generalize. In large measure this is because mental events are not directly observable; they must be inferred. This is the basis for the ancient mind-body dilemma as well as the basis for the rejection by behavioral scientists in the early 1900s of mental events as legitimate scientific concepts. Such "behaviorism" is out of fashion now. Computer science motivated cognitive science and, by indirect inference, made computer programs admissible representations of what people think. However, the limits on direct measurement of mental events do not seem to go away.

Nevertheless, as researchers, we try our best to capture mental models of objects and events that have a referent counterpart in the observable world (e.g., a mental model of how some physical mechanism works). One does this by eliciting drawings, diagrams and flow graphs, verbal stories, and so on. Elicitation of a mental model remains an art. There is no current consensus on how best to do it, although there is a rich literature on the topic (Johnson-Laird, 1993; Moray, 1997a, 1997b; Rouse & Hammer, 1991; Rouse, Hammer, & Lewis, 1989; Rouse & Morris, 1986).

Teams of researchers have engaged in various multiple-year efforts to capture complex human-machine behavior in models. The efforts at manual control modeling (Kleinman et al., 1970; McRuer & Jex, 1967) were discussed in Chapter 3. Another, called MIDAS, was developed at the NASA Ames Research Center. It emphasized visual attention, memory, and workload in helicopter piloting, with task models to help predict times to initiate various maneuvers (Corker, Pisanich, & Bunzo, 1997). A third, called SOAR, used artificial intelligence rule-based methods to model basic cognitive tasks, such as abstract problem solving (Newall, 1990).

All these models are normative in some sense — that is, they are based on some logical rules about how an operator should respond. Most engineering-oriented modelers accept the normative basis and would test how well human behavior abides by these norms. Others, however, primarily coming from the cognitive psychology side, regard the use of normative models as the wrong way to characterize human behavior (Kahneman, Slovic, & Tversky, 1982; Klein, 1989). Bainbridge (1997) asserted that the following aspects of human behavior need to be represented:

- an overview buildup of the operator's current understanding of the situation and what to do about it;
- flexible, adaptable sequencing behavior for handling multiple subtasks; and
- dealing with unfamiliarity in problem solving.

Unfortunately, there is a dearth of human behavioral models that are not normative but that at the same time lend themselves to engineering design.

Mental Model and Decision Aid: What Is the Causality?

Supervisors of automation, some allege, employ their own internal mental model and may, in addition, use an external decision aid. If operators choose to follow the aid's advice, what can be said about their exercising of their own mental model? If they claim to "know" certain facts or have a plan that happens to correspond to the knowledge or plan offered by the decision aid, can the decision aid be considered a convenient normative yardstick against which to measure inferred mental activity? Where there is free will and free interaction between the two, it is difficult to determine what is mentally caused and what is computer caused. Mental and computer models might more easily be treated as a combined entity that produces operator decisions, in which the causal source is not known and is not sought.

This is not unlike the Gibson-Heidegger (Gibson, 1979) perspective on perception to which I alluded earlier. Behavior affects the environment, and the environment affects behavior. In fact, whenever there is causal coupling in physical interactions (e.g., an *energy couple,* such as force on and velocity of a spring, viscous damper, or mass; or voltage across and current through an electrical impedance), it is difficult to say which is cause and which is effect.

What to Do in the Model with the Operator's Free Will?

Most people believe that human operators of machines have free will. The tendency toward greater complexity and robustness in automation technology, as compared with older human-machine systems, and the changed role of the human in becoming a supervisor call particular attention to human free will, seeming almost to defy closed-form modeling. How can free will be captured in a predictive model?

To rein in this unpredictability, the designer may feel obliged to constrain the operator's will, to explicate goals and trade-offs between relative worths of decision alternatives, and to embody these in rigid procedures and computer-based support systems to advise the operator. When the designer does feel confident to constrain and advise the human, reasons should be presented in an understandable and convincing way. Otherwise the operator is likely to assume that design decisions and procedures are arbitrary and may try to overcome or second-guess the design and procedures.

Limits on Design of Human-Automation Systems

The Most Shortsighted Approach

Figure 8.4 portrays a procrustean bed. A legendary ancient Greek tyrant named Procrustes had his houseguests stretched to fit the available beds. Engineers — in fact, designers and managers in general — have often taken the procrustean approach to the human with respect to their designs: Let them adapt; with sufficient training and effort, the operators will be able to cope. Human factors engineers spend a good deal of their time convincing design engineers about the perils of this approach.

Simpler, Easier, and Safer: Does Automation Help or Hinder?

One motivation behind automation is to make the system simpler and cheaper to engineer. A second is to relieve the human operator, to reduce the mental workload. A third, and probably the most important, is to make the system safer. As noted earlier, automation can have just the opposite effect in all three categories (Bainbridge, 1983).

It is usually desirable for the operator to have some idea what the automation is doing and, if the automation fails in any way, for the operator to know how it failed. So, as noted earlier, it is sometimes necessary to provide alarms for all the various subsystems. As systems become more complex, that means more alarms. In some complex systems, such as nuclear power plants, the alarms number in the thousands. Then, as occurred at Three Mile Island, when an accident does occur, the control panel lights up like a Christmas tree and the operator has no idea what is going on.

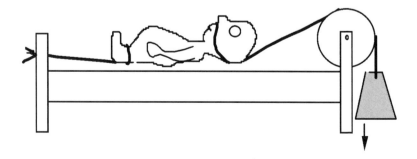

Figure 8.4. The procrustean bed.

Smart computer-based systems are likely to be opaque to operators. It is not easy for the computer to reveal its intentions to the operator. Aircraft pilots express a common complaint about the complexity of modern glass cockpits and computer-based flight management systems: "What is the automation doing now?" they have been heard to ask. "What will it do next?"

An aspect of the technological imperative is to provide flexibility. As discussed in previous chapters, it has seemed to engineers a good idea to allow the operator flexibility in the way information is presented on displays by offering different display modes. It has seemed sensible to allow the operator a variety of control modes so he or she can select which is best to suit the present circumstances. As noted, however, operators can easily forget which display or control mode the system is in. What appears to be happening based on the assumption of one mode may be a deadly delusion if the system has actually been set into a different mode. The so-called mode error is a problem that has been the source of more than one multifatality accident. The only answer, it would seem, is to proceed with automation only after exercising great caution and circumspection about the kinds of pitfalls alluded to here.

Design of Human-Automation Systems Is "Different" Because of the Human

The design of human-automation systems is different from the design of hardware or software in several ways:

1. Design engineers typically have some understanding of physics and can empathize with mechanical and electrical forces and geometry. Seldom, however, do they have much understanding of psychology and physiology or much empathy for the human users who do not know engineering. When given the task of designing human-machine systems, design engineers have to be taught that the object is not design of a thing but design of a relationship between a human and a thing.
2. It is obvious, but important to emphasize, that the human is already "designed." Humans can be selected and trained, but otherwise their limits are to be reckoned with as they are.
3. The human brain does not work according to crisp rules. The rules are fuzzy, in the sense of fuzzy logic. Words and pictures are the brain's normal medium of communication. Words and pictures have imprecise meanings and associations with many other words and pictures.
4. There are many possible modes and channels of communication between human and machine, and redundancy is typically a good thing. A single communication channel may be insufficiently reliable.
5. Humans are able to self-pace in performing tasks, speeding up on the easy parts and slowing down on the hard parts. Automation normally cannot do this.

Humans are prone to making numerous errors, but we are also good at discovering our own errors and correcting them in time and to such a degree that the errors seldom have serious consequences. Humans must be free to make small errors to experiment, to adjust their own behavior in an effort to learn and improve. Humans should never be expected to perform tasks repetitively in the same way and with zero error.

Humans are attributed with the properties of creativity, free will, and indeterminism. Generally it is desired that human operators retain these human traits and apply them with discretion on the job, not think and act like robots. To the degree that human operators exhibit these desirable traits, the behavior of human-automation systems is not so predictable.

Chapter 9

SOCIAL ISSUES OF HUMANS AND AUTOMATION

Social Benefits of Automation

The Obvious Advantages of Automation

Automation has obvious advantages over humans, which are usually the reasons given for implementing it in any given context:

Information-processing speed. Modern computers compute at 1 GHz or greater. Human deductive capability appears to be much slower, but a general number is impossible to fix. For complex pattern recognition and reasoning, the difference may not be so great.

Movement speed. Modern robots respond at 1 kHz, whereas humans respond at 1 Hz, 1000 times slower.

Mechanical power. A robot can lift and carry many tons, whereas a human can handle no more than about 45 kg (100 pounds).

Accuracy and precision. A machine can work at the atomic scale, whereas a human has barely enough visual acuity and dexterity to thread a needle.

Duration of effort. Both information and mechanical automation can work steadily until breakdown from wear. A human loses attention after 30 min of monitoring and, in any case, needs to take breaks for eating and sleeping.

Robustness to environmental conditions. Automation is far more robust than the human with respect to both high and low temperature, high and low pressure, vibration, and ionizing radiation. The human can endure only within a relatively narrow range of these variables without detriment to health.

Reliability. Performance with respect to all these variables makes automation more reliable, at least within the nominal range of its capabilities.

Dollar cost. Automation is usually cheaper than a human in those tasks within the range of machine capabilities, for all the reasons mentioned. This includes consideration of the peripheral infrastructure required to maintain the machine (routine maintenance and repair, etc.) as compared with that required for human labor (food, housing, toilet facilities, medical benefits, insurance, recreation, etc.).

Enabling of Remote Control by Humans

Modern broadband communication means that automation can be put to work anywhere and controlled from anywhere. Computers can now be accessed over the Internet from any terminal to provide whatever information they have been programmed to provide. Air traffic control centers and electrical power dispatching centers need not be located near where the end operations are performed but can be located at sites most convenient from the standpoint of land cost, housing, or transportation for employees.

However, remote control of mechanical devices that perform physical tasks is usually limited to authorized operators. This is true of pilots of remotely controlled air, land, or undersea vehicles; of train dispatchers who operate track switches from centralized control facilities; and of operators in control rooms for manufacturing or process control plants who activate motors or sensors at various locations out in the plants.

Soon, however, the Internet may be used to control physical tasks. It may be difficult today to imagine why one might wish to exercise remote control for heating, cooling, or lighting the home; for cooking, cleaning, watering the lawn, security, and so forth — but surely the capability is there, and more efforts to market such capability will be seen. Perhaps the desire to remotely *monitor* the state of many variables in the home is easier to comprehend. Perhaps more interesting is monitoring other people (children, elderly persons) or pets. Remote monitoring of people raises a special set of ethical issues concerning privacy.

The ability of head-mounted visual displays and their auditory and haptic counterparts to make people feel telepresent at some location other than where they really are, or to allow people to experience presence in a computer-programmed virtual reality, has resurrected old philosophical questions of *being* and *existence* in the field of ontology (Sheridan, 1999).

Understanding Ourselves by Emulating Ourselves

Research in automation and its derivative engineering disciplines, such as artificial intelligence (AI), biomedical engineering, cybernetics, and systems engineering generally, have included many efforts to emulate human and other animal behavior — in sensing, walking, manipulating, and information processing (memory and recall, pattern recognition, decision making). The challenge of building and programming robots is in part motivated by a latent desire of humankind to emulate itself. Though emulating animal behavior is not always the best technological solution (e.g., modern airplanes do not fly like birds), these efforts do lead to a deeper understanding of humans by humans.

The Turing test for intelligence (named after the British mathematician who invented the Turing machine, a logical analysis of the nature of computers as we know them today) has evolved into a popular annual contest. Computers with different question-answering software compete against each other to try to fool human judges, who freely ask questions (typewritten through a computer network) and judge whether the answerer is a person or a computer.

A current trend in AI is *embodied intelligence*, referring to the capability of robots to appear to have human mannerisms (e.g., with their eyes and body language in responding to human actions in a way that seems humanlike).

These efforts to emulate and understand ourselves are bound to have important long-term effects on science, ethics, religion, and culture in general.

Cathedral Building

In ancient times communities of people labored over long periods to construct cathedrals, temples, pyramids, and other sacred structures to honor their deities and prove their devotion to the greater cause. The results were magnificent feats of technology. Notwithstanding that in many cases the labor was involuntary, these communities did feel pride in the results.

Those efforts might be likened to modern-day space probes, robots, computer chess players, and other high-tech achievements that are built to demonstrate scientific prowess rather than to solve pressing human problems or to sell to commercial markets. Just as the building of cathedrals and temples in the Middle Ages posed technological challenges and demonstrated technical achievement, so too do the modern feats of automation yield community satisfaction in the results. Just as the cathedrals were supposedly built to glorify God (and perhaps a few lesser leaders of the day claimed some attention in the process), the modern technological cathedrals may be said to glorify the scientific theories (with the scientists and engineers hoping to share the glory).

Human Irrationality

Infatuation with Technology

Some people delude themselves into thinking that more capability for control necessarily means better system performance. But anyone who has struggled with an unfamiliar VCR remote control knows otherwise. Still, manufacturers of home appliances play to this human frailty by adding relatively useless "features" to otherwise sensible designs and charging more money. One artist portrays this situation by showing a naked fellow helplessly suspended by a rope inside a towering silo, the walls of which are covered with sophisticated-looking dials, knobs, and buttons. The man's smile suggests a feeling of great pleasure and confidence and a sense of great power. Maybe the knobs, dials, and buttons are not connected to anything, but that may not matter. As a small child one may derive pleasure in make-believe.

Factors that Bias Decision Makers

In Chapter 4, I discussed biases of decision makers, especially with regard to how probability judgments tend to deviate from what is predicted by classical normative decision theory — for example, Bayesian updating (see Appendix A). Researchers have proposed many reasons, some more clearly understood than others. Human-automation system designers should be aware of them. Other factors that affect human decision making are not explicitly biases in probability estimation. Some such factors are the following.

People are unable to keep multiple hypotheses in their minds; they are not good at aggregating data. This may explain why they tend to hedge their bets toward the prior expectation or the mean of the range. In light of the brain's "computational noise," this might be considered rational.

People tend to regard situations as more risky if persons other than themselves are in control, and especially if they are coerced into such situations. People also tend to trust themselves or other people more than they do automation, even though the automation may be reliable. This would be rational behavior if they had no knowledge of the other person's or the automation's capabilities and were risk-averse.

In comparing humans with machines (in determining allocation of functions between humans and machines and assessing reliability of human-machine systems), there is a tendency to regard people as having a few simple failure modes, much as do machines. In fact, humans differ enormously from machines, in that they are inherently variable and unreliable in their detailed behavior, while simultaneously being hyperadaptable and metastable in their overall behavior because they perceive and correct their own errors. Thus they have uncountably many failure modes and are not amenable to being characterized by simple reliability numbers.

People are hedonistic in their decisions. That is, they favor benefit to themselves, their own loved ones, social class, community (nation, race, etc.), and generation over benefit to persons who are removed with respect to these factors. This is certainly what one would expect, but it is not according to ideals of social equity, morality, and religious altruism. Nor is it in the best interest of propagation of the human species. Loss of a peasant's life to AIDS or famine in a third-world African village and loss of the life of an astronaut, a corporate chief executive, or a president in a first-world nation tend to be evaluated differently by many orders of magnitude, at least in terms of the money allocated to protect against such loss. Although this is a bias, from purely economic criteria, some might judge it to be rational.

Most people exhibit a lack of understanding of the basic concepts of decision science, such as probability and utility. Further, they often seem to lack an evidentiary basis for their own beliefs, but they are willing to accept hearsay, fashion, and isolated sensational anecdotes in forming their beliefs.

An Impediment of Democracy: Lack of Standardization

In some Western nations the principles of democracy can lead to irrationality. An example is found in the construction of nuclear power plants in the United States. Every U.S. nuclear power plant is essentially different because architect-engineers have selected, cafeteria style, from a wide array of vendors for each piece of equipment, making up a unique hodgepodge combination for each plant. Even in plants located immediately next to each other and operated by the same management, such as the two on Pennsylvania's Three Mile Island, the combination is different from one plant to the other. This contrasts with France, where nuclear plants are standardized.

Variety is fine for homes, clothing, or other areas of life in which variety and self-expression add to the culture. However, for nuclear power plants, transport aircraft, or automobiles, where safety and predictability are essential, variety is wasteful in cost and time and hinders understanding of safety and reliability. Standardization has proven to lead to greater safety and cost-effectiveness.

Quantitative Modeling as Sorcery for the Powerful

Technologists consider it their business to quantify things. They are hired to do this. Quantification is also their identity, their pleasure, and their reward, so they are naturally motivated to quantify whenever the opportunity presents itself. I include myself in this group. The clear and present danger is to overdo it.

Mathematical models are sometimes paraded before politicians to present a facade of objectivity. This may be done with full knowledge that the audience is incompetent to appreciate the limitations of the models or to ask meaningful questions about their validity. Sometimes this is called "lying with statistics." One can liken this practice to the kings of old employing court sorcerers, who used incense and uttered strange incantations to frighten royal visitors and ensure the king's power.

Some nations, the United States being a prime example, seem compelled to put numbers on safety of nuclear plants and other complex regulated systems of humans and machines when public safety is of concern. Sometimes the numbers have little validity. Other nations (I experienced expressions by Japanese officials of this viewpoint) seem satisfied that if the people responsible for system safety are trying their hardest and doing their best with available resources, that is good enough, and therefore their system is acceptably safe.

The obligation for the technically literate, it seems to me, is to help bring rationality and truth to all discussions of model predictability, especially as concerns safety and security. One hundred percent predictability or safety, which some politicians pompously and irresponsibly demand, is unattainable; it is possible, however, to continue to make models and systems increasingly better. Absolute reliability in predictions about relatively new or changing technology or complex human-machine interactions, for which only the most primitive analytical tools are available, make little sense. What is more believable, in my judgment, is sensitivity analysis, which assumes and admits to some rough reliability number, then analyzes whether that number increases or decreases as a function of certain actions that are taken to configure or to control the system.

For some kinds of technology, reliability modeling seems almost infeasible. One particularly worth mentioning is computer reliability. Not so many years ago, I experienced how, when planning intelligent transportation systems, automotive engineers (who fully appreciated reliability problems with brakes and concrete structures) had the idea that once a computer program was compiled, it was guaranteed to be reliable. They did not appreciate that the number of failure modes of modern computers interacting with external environments, especially with human users, is virtually infinite — and can never be exhaustively evaluated.

Advocacy versus Objectivity

Scientists and engineers like to claim objectivity. Perhaps that is the reason many interests of human users and human operators are neglected — the designer simply hasn't found an objective way to accommodate the "soft" considerations. So those considerations are ignored, and the effort is made to "optimize" on the basis of only those variables that can be quantized.

However, such a practice is suboptimal. Just because a variable cannot be quantized does not mean it is unimportant; indeed, if it is left out of consideration, the system may not be optimized.

In contrast to science, Western jurisprudence operates not on the basis of objectivity but on the basis of advocacy — advocates for various sides of a contest try to convince judges or juries. Everyone knows that real decision making in technological design and implementation involves advocacy at many levels. Unfortunately, the human user or operator has often been left without an advocate. This is changing. The human factors profession, while trying for objectivity when it is feasible, has taken to advocacy when worthwhile considerations do not lend to objectivity. So have unions, government regulators, and consumer groups. The need is for a greater understanding of when objectivity can be employed and when a fair process of advocacy should be invoked.

Peer Influence

The magnitude of peer social influence on people of all ages is well established. Although people often decry the extent of peer influence on children, we allow ourselves to be influenced by the media and by our own local cultures much more than we might admit.

A simple and oft-repeated psychophysical experiment is illustrative. Two parallel lines of slightly different length are drawn on a piece of paper. The experimental participant is allowed to get close enough so that without any external influence, he or she would almost always be able to select which is the longer of the two. However, the experimenter leaks a false report to the participant indicating that the majority of participants selected as the longer line the one that is actually shorter. Experimental results almost always show that participants so treated are heavily influenced by the leaked data and therefore make the wrong selection — even though their direct visual perception tells them otherwise.

Safety: Normative Objectivity versus Media-Motivated Acceptability

Safety, to a risk analyst (*normative* safety), is the expected value of avoiding bad consequences, where weighting of different consequences (apart from probability) is based on some combination of human suffering and economic loss. In the real world, what is safe is what is judged to be acceptably safe. Social belief and social acceptability seem to be the ultimate criteria of safety, quite apart from the product of probabilities and consequences — but what underlies social acceptability?

There is something about sensational accident situations that gives rise to far more importance weighting than is warranted, at least in comparison with mundane accident situations. A commercial airplane accident is picked up by the wire services and makes headlines everywhere. It is sensational because it is unexpected (and therefore, in an information theory sense, amounts to lots of information). An isolated single- or multiple-fatality automobile accident often will only make the back pages of a local newspaper. It is expected and accepted that there will be many automobile accidents and that some will result in loss of life, so there is not much information there. Somehow, the seriousness of an accident, like the seriousness of any public event, is judged according to the extent of its media coverage, which in turn is truly dependent on its unexpectedness or information magnitude, apart from the extent of deaths, injuries, or property damage.

So, does the multiple-fatality accident garner media importance out of proportion to the number of fatalities? Not always. Strangely, some people see it the other way around. I recall executives of a particular non-U.S. national rail system expressing the opinion that a train wreck in which N people are killed is not equivalent to N train wrecks in each of which one person is killed. They saw all train wrecks as almost equally bad — in terms of media coverage. Maybe the public empathizes as much with the size of the physical wreckage as it does with the number of fatalities!

Alienation of the Individual

The term *alienation* is used in this section to characterize a variety of negative impacts of automation on the individual human. When systems are automated and the human is removed not only spatially but also temporally, functionally, and cognitively from the ongoing physical process he or she is directing, that can be called *alienation*. This is because what the human does and thinks are likely to have different timing, different explicit form, and different logical content from what the automation does and thinks. The machine's behavior is alien to the human's.

People who have been pushed sideways into jobs that do not make use of the skills on which they pride themselves and with which they identify will feel frustrated and resentful. People who no longer understand the basis of what they are asked to do will become confused and distrustful. Especially if they perceive a powerful computer to be mediating between them and what is being produced at the other end, people become mystified about how things work, question their own influence and accountability, and ultimately abandon their sense of responsibility. All this can lead to alienation (Sheridan, 1980; Zuboff, 1988). The components of potential alienation are detailed in the following sections.

Threatened or Actual Unemployment

This is the factor that has been most often cited in opposition to automation. In the English knitting mills of the mid-19th century, angry mobs, allegedly led by "King Ludd," rioted and smashed the newly "automated" machines out of fear of losing their jobs. They came to be known as *Luddites*, and the term has come to be used against any group opposing the advance of technology, especially out of fear of unemployment.

For many years the advance of automation seemed to have little significant impact on unemployment, because people replaced by automation were simply moved to other jobs, such as machine supervision and maintenance. However, organizations have become more efficient in their use of human supervisors: Fewer people can now supervise more machines. Furthermore, automation is becoming more able to detect its own failures, and in some cases it can even repair itself. The real threat today is unemployment of the unskilled and technologically illiterate.

Erratic Mental Workload and Work Dissatisfaction

Automation affects not only the nature but also the pace of work, and it makes that pace vary between extremes. Airline pilots and nuclear plant operators refer to their work as "hours of boredom punctuated by moments of terror." Some workers may feel that although automation has eliminated many mundane tasks from their work, it has not relieved the stress induced by on-the-job decisions. For example, some pilots feel that in times of emergency or high stress, the automation is not sufficient to ensure safety and that even more human involvement is needed than in earlier times of pure manual control.

Although manufacturing workers are often glad that automation has broken the dull monotony of repetitive manual work, they may not get as much satisfaction from their new supervisory monitoring role. The satisfaction derived from putting in a hard day's work and producing something with their own wits and efforts may be gone.

Already, some airline pilots are complaining about automation, some manufacturing engineers are having second thoughts about robots, and some power plant operators who have experienced older, simpler, and very reliable plants are wondering if things haven't gone too far.

Centralization of Management Control and Loss of Worker Control

One result of automation and the introduction of electronic technology often feared by workers is the possibility that management can secretly record and monitor their work output through the equipment's sensors and communication capabilities. Initially managers assured workers that they did not engage in such activities, but today it is clear that productivity data are so easy to record that they are being recorded. Such monitoring is occurring not only in industrial plants but also in hospitals, offices, trucks, and aircraft, where high-bandwidth communication is becoming plentiful and makes such monitoring from a central location easy.

The mere possibility of being monitored in this way is often sufficient to produce worker anxiety, including fear that private data stored electronically may be accessed by persons other than those authorized.

Centralization of monitoring and control, so easy with advanced automation, is often seen by management as synonymous with efficiency. In some cases centralization may enhance productivity, but in other cases it may prove detrimental.

Desocialization

Although cockpits and control rooms now require teams of two or three, the trend is toward fewer people per team, and eventually one person will be adequate in many workplaces. This is certainly true in factories and offices. Thus cognitive interaction with computers is gradually replacing that with other people. As supervisory control systems are interconnected, the computer will mediate increasingly more of what interpersonal contact remains, as has already happened in many cases with e-mail, pagers, and associated software for management coordination and computer-supported cooperative work. Although interpersonal communication is in one sense becoming easier, in another sense it is becoming artificial (Turkle, 1995).

Deskilling

Skilled workers "promoted" to supervisory controllers (sometimes derogatorily referred to as "button pushers") may resent the transition. In part, this may be out of fear that when called on to take over and do the job manually, they may not be able to do so. A skill such as manual machining, developed over a long period, provides the worker a sense of dignity and self-respect. The same may be said for the seamstress or other manual artisan. If the skilled worker becomes a button-pushing supervisor and monitor, that original sensory-motor skill may atrophy. Many researchers have decried the tendency for skill to atrophy when automation is used (Bainbridge, 1987; Kaber & Riley, 1999; Wiener & Curry, 1980).

Intimidation of Greater Power

Automation and supervisory control encourage larger aggregations of interconnected equipment, higher speeds, greater complexity, and probably greater economic risk if something goes wrong and the supervisor doesn't take the appropriate corrective action. The human supervisor will be forced to assume increasingly more ultimate responsibility (although in most cases the responsibility probably should reside with some combination of the manager and the system designer). Depending on one's personality, this could be very intimidating and lead to stress, inattention to detail, anxiety that one is not up to the job's requirements, or paranoia.

Technological Illiteracy

In older manual control systems, the operator could understand how things worked. In the role of supervisory controller, the operator may lack the technological understanding of how the computer and the rest of the complex technology do what they do. What is really going on with the communications and control software may be too specialized even for most technicians involved with the newer systems. The automobile mechanic's inability to understand the computer/electronic systems in modern automobiles is an example. Operators may come to resent this and to resent the elite class of technologists who do understand.

Mystification and Misplaced Trust

Human operators of computer-based systems sometimes become mystified by, and superstitious about, the power of the computer, even seeing it as a kind of magic or a "big brother" authority figure. This leads naturally to naive and misplaced trust. This was particularly well articulated in Norbert Wiener's (1964) *God and Golem, Incorporated*, in which he used a classic in horror literature, W. W. Jacobs's *The Monkey's Paw*, as a metaphor:

"In this tale, an English working family sits down to dinner in its kitchen. The son leaves to work at a factory, and the old parents listen to the tales of their guest, a sergeant major back from service in the Indian army. He tells of Indian magic and shows them a dried monkey's paw which, he tells them, is a talisman that has been endowed by an Indian holy man with the virtue of giving three wishes to each of three successive owners. This, he says, was to prove the folly of defying fate.

"He claims he does not know what were the first two wishes of the first owner, but that the last one was for death. He himself was the second owner, but his experiences were too terrible to relate. He is about to cast the paw on the coal fire when his host retrieves it, and despite all the sergeant-major can do, wishes for £200.

"Shortly thereafter there is a knock at the door. A very solemn gentleman is there from the company that has employed his son. As gently as he can, he breaks the news that the son has been killed in an accident at the factory. Without recognizing any responsibility in the matter, the company offers its sympathy and £200 as a solatium . . .

"The theme (here) is the danger of magic. This seems to lie in the fact that the operation of magic is singularly literal-minded, and that if it grants you anything at all it grants what you asked for, not what you should have asked for or what you intend. If you ask for £200 and do not express the condition that you do not wish it at the cost of the life of your son, £200 you will get whether your son lives or dies.

"The magic of automation, and in particular the magic of automatization in which the devices learn, may be expected to be similarly literal-minded" (p. 58).

To a naive user the computer can be simultaneously so wonderful as to seem faultless, and if the computer produces other than what its user expects, that can be attributed to its superior wisdom. Such discrepancies are usually harmless, but if they are allowed to continue they can, in some complex and highly interconnected systems, endanger lives. It is therefore crucial, as new computer and control technology is introduced, that it come to be accepted by users for what it is — a tool meant to serve and be controlled ultimately by human beings.

Sense of Not Contributing

Though the efficiency and mechanical productivity of a new supervisory control system may far exceed that of an earlier manually controlled system, the operator may come to feel that with automation he or she is no longer the source of value added, no longer a significant contributor. This sense of deprivation has been expressed by factory workers, farmers, craftspeople, accountants, and others who feel threatened by automation. The sense of personal productivity — what psychologist Erich Fromm (1995) called the *productive orientation* — is allegedly fundamental to humans' sense of self-worth. Without it, who are we?

Abandonment of Responsibility

As a result of the factors just described, human supervisors of automation may eventually feel they are no longer responsible for what happens but that the computers are. A worker with his or her own set of hand tools or a simple self-powered but manually controlled machine, though sometimes placing the blame for difficulties elsewhere, has a clear responsibility for use and maintenance of the tools or machine. That worker is accountable for what is produced, and everyone knows it.

When workers' actions in using a machine are mediated by a powerful computer, however, the lines of responsibility are not so clear, and the workers may not be sure which should get the credit or the blame for a situation — the computer or themselves. As a result these workers may, in effect, abandon their responsibility for the task performed or the good produced, believing instead that it is in the "hands" of the computer. Even when computers are installed to aid information flow from one worker to another, or to act as a processor or storer of data, individuals using the system may feel that the machine is in complete control and disclaim personal accountability for any error or performance degradation.

Blissful Enslavement

To many writers the worst form of alienation, the worst tragedy, occurs when a worker is happy to accept a role in which he or she is made to feel powerful but, in actuality, he or she is enslaved. Engelberger (1981), the founder of the world's first major robot company, reminded readers that "it will always be far easier to make a robot of a man rather than to make a robot like a man." Both Huxley's *Brave New World* and Orwell's *1984* are famous for the theme of blissful enslavement.

With regard to the affects of the computer on a person's self-perception, I am fond of citing historian Mazlish (1967), who referred to the computer as the "fourth discontinuity" in this self-perception. At first, said Mazlish, humans saw themselves as the center of all things, but Copernicus jarred this perception by showing that the human race was an isolated dweller of a tiny planet of a minor star (which turned out to be at the edge of only one galaxy). Darwin came along and rudely asserted that we were descended from the apes and they, from lower creatures still. Later Freud suggested that we humans are not even consciously in charge of our own faculties, that our egos and ids drive us. Now the computer may be dealing us the ultimate affront — surpassing us intellectually — beating us in our best suit, and seemingly inviting us to give in.

Surely automation promises greater product quality, better energy and economic efficiency, and improved worker safety — all benefits that motivate development of the new technology. Nevertheless, these potential negative impacts must be examined and reduced in order to achieve a satisfactory acceptance of automation by society.

Alienation of the Community

Automation can be alienating not only to individuals but also to communities of people as groups. Using that somewhat arbitrary dichotomy, this section considers additional forms of automation alienation.

Distributive Justice: Who Decides and Who Benefits

Automation in any particular application has its proponents and its opponents. The proponents are usually the technologists who design, build, market, install, and manage it. They are the ones who usually benefit directly from its use. The operators can be opponents, as discussed in the previous section, or proponents, seeing themselves as its caretakers. Others, who have no direct participation in it, may be threatened by various aspects of it, because of economic competition, because they are jealous or intimidated, or for other reasons. In a democracy it is an individual's or a corporate entity's right to automate, so long as legal constraints on safety and environmental impacts are met. However, as with any human endeavor that affects many people, there is bound to be disagreement over who gains and who loses, who decides and who has the decision forced upon them. These are issues of *distributive justice* (Rawls, 1999), and they are not easy.

Figure 9.1 illustrates the problem of distributive justice that occurs when rich nations or individuals have the power to automate and poor nations or individuals do not.

Questions of distributive justice occur at two levels. At the global level, implied by the preceding paragraph, all affected parties are included. However, there is a lower level having to do with the allocation of power among system designer, system manager, and human operator of the automation. There is a tension between the prescriptions and proscriptions explicitly specified by the designer (or implicit in the system design constraints), the policies and constraints established by system management, and the freedom of the operator (free will) to make and execute decisions in the course of system operation. How can this tension be resolved?

Technophiles and Technophobes

Perhaps it is good that some people are *technophiles* and others are *technophobes*. The technophiles want to exercise technology wherever possible, seeing it as a challenge. However, they can succumb to the temptations of promoting the technology for self-gain at the expense of others. The technophobes see technology as a threat to tradition and to well-established and comfortable ways of doing business. They might have society return to what they believe would be a safer, healthier world without technology, modern travel and commerce, and large-scale interconnected communications, power, food, and water supply systems that are vulnerable to terrorists. However, one cannot escape the fact that indi-

the First
and Third Worlds
are one and
the same system

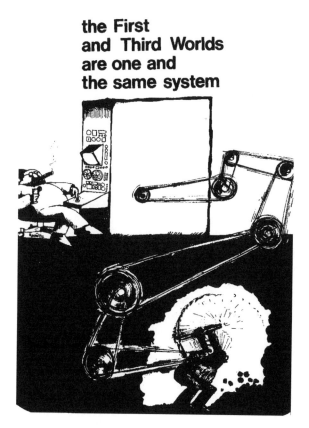

Figure 9.1 A political cartoon by Claudius (Brazil). Used with permission.

vidual longevity and health are far better than they were in earlier times, and in any case the clock cannot be turned back.

Many find themselves between the extremes, maintaining a neutral stance. Sometimes they are conscious of the evidence tugging them in two directions. Sometimes their neutrality is based on ignorance or apathy. Although the media may provoke a war of words between the two extremes, the public is often confused by the arguments and is made to feel left out. Nevertheless, the encounter between technophiles and techno-phobes should be useful for elucidating the truth and directing compromise toward sanity in public policy.

Machine Productivity instead of Human Productivity
Today national governments and multinational corporations are obsessed with productivity. Balance of payments and national well-being

are at stake. Computers integrated with sensors, actuators, and robots are allegedly there to make the factories, power and chemical plants, aircraft, ships, automobiles, banks, hospitals, schools, and homes more productive. Most people consider economic productivity to be the overriding social goal.

What kind of productivity is being increased? Is it the kind of productivity that at the end of the day — and indeed at the end of our lives — gives us the satisfaction of having best served other human beings? Is machine productivity synonymous with human productivity? Or are the two different — perhaps orthogonal, or even in conflict with one another?

Daily Living by Remote Control: Reduction of Social Contact

Future constraints on energy potentially will limit free travel. However, with microelectronics advancing communication in so many ways, people have a new capability to use telephones, fax machines, desktop computer terminals, and portable communication devices to order fast food or supplies delivered to our homes, to shop from electronic catalogs, to buy and sell stocks, to pay bills or make bank transfers, and to do e-business without leaving home. At the moment these personal remote control activities are not regarded the same as remote control of industrial automation that does useful mechanical work but, rather, as an extension of the telephone for communication purposes.

Future scenarios may include remote control of robots — to browse up and down the aisles of the supermarket or store to inspect and handle merchandise through remote video links, to cut the grass, or to take out the trash. It is too early to predict the benefits and hazards of such radical changes in living style. However, no significant breakthroughs in technology are required for such a scenario to occur.

The tendency of people to perform many more daily functions by remote control requires fundamental rethinking of the meaning of home, commerce, transportation, education, and human fulfillment. It probably spells a significant reduction of casual social contact with others in our communities. Our social contacts may have to become planned and controlled to a much greater extent. Does this translate to psychological or political dangers lurking ahead?

Electronic Telegovernance by the Powerful over the Powerless

The use of modern automated communication technology can provide not only feed-forward of information from those in power down to the masses of citizens but also feedback from the masses to the powerful

(DeSola Pool, 1973; Sheridan, 1975). Many experiments have been conducted with people watching television programs and "talking back" through the telephone or the Internet to vote on questions posed. Conceivably, given the bandwidth already available, democratic governance could extend to all citizens and be essentially instantaneous, possibly even eliminating layers of elected representatives.

This vision may have advantages for some purposes — for example, to educate and test public opinion on certain issues. However, such possibilities of electronic governance also conjure up visions of a society that could rapidly put too much power in the hands of whoever controlled the head end of the communication system. Indeed, such arguments have already been made against the national media networks, and similar arguments underlay the motivations for breaking up the AT&T and (alleged) Microsoft monopolies.

Automation of Ecoexploitation

The year 1970 saw the beginning of great international debate among scholars about "limits to growth" hosted by the Club of Rome and other international study organizations (Oltmans, 1974). The issue of exploitation of natural resources was central, and the role of automation was a great concern. Some assert that the debate fizzled and came to naught, whereas others claim there were lasting revelations and lessons learned. Most planners of future automation systems believe it is especially important that the so-called technological imperative — that automatic systems be built because they can be built — be restrained, especially with respect to use of energy and other precious natural resources. Many nations are lucky enough to have many such resources. Others are not so lucky.

Proper use of automation will produce the same or greater benefits for people while reducing demands on energy and other (especially nonrenewable) resources. However, automation makes it easy to consume energy and other resources for questionable ends. It would be criminal to employ automation that, for the sake of short-term productivity, gluttonously exhausts nonrenewable resources and leaves none for less-developed nations and for future generations. For example, certain wealthy nations now recover cobalt and other mineral nodules from the ocean bottom using high-tech robotic suction devices. In some cases this happens just offshore of poorer nations that are powerless to stop it.

Society may be said to be facing a new stage of Hardin's (1968) oft-quoted "tragedy of the commons." English sheepherders unintentionally increased the number of sheep beyond what the common grazing lands could support by each adding a number that was (to the sheepherder) insignificant. Now the technologically advanced nations are automating and expanding their use of automatic industrial processes that foul the air and water and use up the oil and other natural resources. Would global warming progress as fast if there were no automation? The metaphor of the sorcerer's apprentice in Walt Disney's *Fantasia* also comes to mind: an unstable self-multiplication of trouble.

Telerobotic Soldiers, Spies, and Saboteurs

Consider a future manually controlled teleoperator that has good mobility on land, in the sea, or in the air or in space — all are possible, all are being developed. Assume wireless telecommunication, reasonable sensors, and some manipulative dexterity (full telepresence is not necessary). Assume a good battery or other power supply, and the possibility that the device can be controlled from afar to "plug itself in" from time to time to reenergize itself. This teleoperator may be driven by its human operator to do mischief in anyone's backyard. Its human operator need have no empathy for its welfare, for it is only a mechanical slave. Further, at any time its human operator can cease to communicate with it and with ease can abandon responsibility and accountability for its behavior.

Now consider a future telerobot, in this case with a computer to help it see, hear, and touch, but also having some motor reflex skills to conserve energy and adapt to its environment and having enough knowledge and instruction to understand how to implement its programmed goals in spite of disturbances or obstacles it encounters. This telerobot may be sent to do even more mischief because it can go surreptitiously or lie dormant for long periods without the need for communication. It can be used to spy, sabotage, set explosives, and perform a host of other duties. Because it is controlled in supervisory fashion, any one human supervisor can multiply his or her own capability manyfold. In this case it is even easier for human operators to abandon responsibility. They can claim to be too busy with their brood of telerobots, and in any case each telerobot can be said to have a mind of its own.

In past arguments, fights, and wars, individuals who felt they must resort to violence toward other individuals put themselves at some bodily risk in doing so — at least, more risk than those not so inclined. Traditionally such behavior has been lauded as courage or bravery and at least posed a high cost on the initiator. However, technology has been changing all that, as weapon telerobots have evolved from clubs to arrows to bullets to bombs dropped from airplanes to smart cruise missiles. The human operators no longer put themselves at risk so much as they jeopardize the safety of others. Telerobot technology is merely the next step.

Some have heralded the dawn of smart weapons, battlefield robotics, and "telegladiators" as a new day when international disputes will be fought on a large technological playing field (space, undersea, or surely somewhere well away from real people). Technological prowess will win the day, and no real people (at least on one's own side) will get hurt. Certain military planners seem to be moving toward that fantasy even now.

Already there are telerobotic sonar spying devices in the oceans and similar electromagnetic spying devices in orbit — all over the protests of the helpless poorer nations. Now space is becoming militarized — perhaps eventually in the form of an ultimate human-machine system programmed in top-down supervisory fashion with the purpose of defending its nation but demanding, ultimately, that people abandon responsibility to the computer. I cannot help but recall again the theme of Wiener's *God and Golem, Incorporated* (1964), mentioned previously in this chapter.

Terrorism and the Vulnerability of a Variety of Automated Systems

Since September 11, 2001, the Western world has realized that many highly automated systems throughout the fabric of society are vulnerable to terrorists and that large systems involving both many people and much capital investment can be shut down through actions that may be relatively simple for hostile agents to execute. Transportation systems, communication systems, postal systems, and supply systems for food, water, electricity, and gas are all highly automated and therefore at risk.

We had already experienced somewhat milder terrorist acts in cyberspace. A number of denial of service attacks have already been effected against commercial Internet sites — the perpetrator secretively planting automatic software agents in the computers of thousands of unsuspecting individuals and synchronizing them to launch a simultaneous attack on their victims. It is extremely difficult to trace the source of such attacks.

Major banks and financial service organizations have been forced to spend many millions of dollars to build firewalls to prevent unauthorized access to their communication networks, which are responsible for secure transfer of billions of dollars per hour. Now the threats to large human-machine systems are far greater. Military and civilian security organizations now freely admit that their own "red teams" (used in war games) have succeeded in invading control systems for electrical power distribution grids, hydroelectric dams, transportation control centers, and even defense facilities, all of which had been assumed to be secure.

Unfortunately, society is entering a new era when we will have to spend large sums of money and human capital to ensure the freedoms and openness that we have experienced in recent years. We may also have to sacrifice some of our freedoms in order to do so.

Chapter 10

SYSTEM MANAGEMENT AND EDUCATION

In this book I have tried to consider the variety of salient research and design issues concerning humans and automation. Research and design, however, do not form a complete story, for they are embedded in a much larger context. The best efforts of individual researchers and designers are not enough. There is still the problem of making the design and research efforts come to some useful fruition to benefit society. This requires management and it requires educating the designers, managers, and public who ultimately must accept the products and services that result from the research and design. In this section some final caveats are made regarding these topics.

System Management

System management is the critical final step in getting research and development of human-interactive automation into use. It involves many considerations, and only a few are mentioned here. The reader is referred to Sage and Rouse (1999) for a full treatment.

Bounding the System

As suggested in Figure 10.1, the operation of human-automation systems depends on their designers. The resources and policies used by designers depend on those available and allocated by their own institutional management. The institutional management depends on government laws and regulations, as well as economic priorities and constraints largely influenced by government. The operation of government in turn is influenced by the prevailing culture, including language, ethical standards, and traditions. Both system designers and managers should be aware of these rings of influence and the successively longer time constants that prevail because the problem to be solved depends on changes further out in the set of rings.

Life Cycle Considerations

Figure 10.2 shows the series of steps (constituting a life cycle) that must be taken in bringing a new system into being. Each step generates the

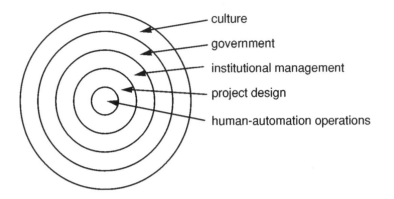

Figure 10.1 The embedding of humans and automation in successively larger contexts.

prerequisites for the next step, and insofar as the necessary conditions are not met (as represented by the feedback arrow at each junction), the predecessor step must be worked further. The output of each step is organized information, and the feedback to the previous step is information discrepancy.

Usually the further down the cascade of steps one goes, the more time consuming and the greater the cost of that step. Scheduling and budgeting must anticipate this fact of life. If an early step — for example, the requirement specification — is not done well, the cost at the later steps is still higher. A requirement should be necessary (not whimsical), should describe what (not how), should be unambiguous, should be complete, should be traceable to the reasons why, should be well documented, should be quantitative, and should be testable (so it is clear if it is fulfilled). These are the requirements of requirements.

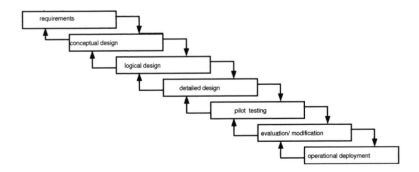

Figure 10.2 System acquisition life cycle (modified from Sage and Rouse, 1999; adapted with permission from Andrew Sage and John Wiley and Sons).

It is essential that human factors professionals work with system engineers at every step of this cascade. If they are brought in only at the end to put their blessing on the design, it is too late for change, and the result is likely to be an unhappy one for all.

It is not necessary that one step be fully completed before the next is begun. To save time there are trends toward *concurrent engineering*, wherein successor steps are begun while predecessor steps are being completed.

Figure 10.3 shows a similar life cycle of stages through which physical material necessarily goes. The output of each stage is a form of the physical material, and the corresponding feedback is a discrepancy, in this case often requiring a new batch of material to satisfy the need when the previous one did not. In selecting materials the systems manager and engineer should consider these stages, though already engineered materials are usually available for the manufacture of the system. The times and costs required for each stage are essential in system planning. Consideration of sustainability and materials costs are increasingly an obligation of the designer or manager, including the whole sequence of events between selection of materials and means of disposal.

Cost Trade-Offs

The costs related to human-automation systems are many and varied. As automation reliability increases, the personnel costs for both operations and maintenance decrease, but the capital cost is higher to gain such

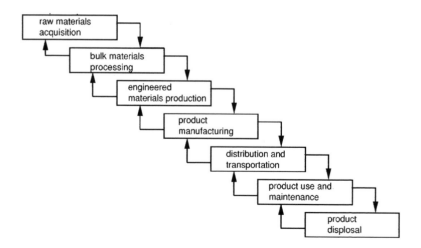

Figure 10.3 Hardware product life-cycle (modified from Sage and Rouse, 1999; adapted with permission from Andrew Sage and John Wiley and Sons).

reliability. However, as system complexity increases, personnel costs may be expected to increase, other factors being equal. If maintainers discard components at the point of failure, the hardware costs increase while repair personnel costs decrease, whereas the opposite is true if components are repaired on site or if effort is expended on preventive maintenance. Standardization usually decreases individual system cost, but there are high personnel and test costs to achieve that standardization. Obviously, cost optimizations can be discovered through quantitative analysis of such cost trade-offs.

It is common to consider the cost of a new system as the hardware acquisition cost. But this is only the tip of the iceberg. Below the surface are the costs for operations, training, software and software support, test, maintenance, repair, retirement, and disposal. The history of U.S. manufacturing plants' acquiring robots is a case in point. In many cases the buyers naively ignored the fact that the infrastructure to support the robots was not in place and not considered very seriously. In the end it greatly outweighed the robot acquisition cost and set back efforts to automate by many years.

Information value theory, as explained briefly in Appendix A, provides a tool for deciding how much should be invested in research to answer particular questions. If the expected design/manufacturing/sales costs saved by knowing the answer do not exceed the costs of doing the research, then there is no point in doing the research.

Human-Automation Systems to Improve System Management

In Chapter 2 I briefly mentioned automation in business systems. The fact is that automation in the process of managing system design is a huge activity, replete with a great variety of computer-aiding and sophisticated software tools. One of the first successful expert systems was used by Digital Equipment Corporation to configure computer systems. Other systems are used for teleconferencing between remote sites, including the sharing of graphical materials, including real-time drawing, between sites. Still other software enables a designer to render 3D drawings or colored solid models.

An example of the latter is software called Jack (by the University of Pennsylvania), which enables the designer to render a human body in a cockpit or other workplace and alter the anthropometric dimensions of the limbs and torso to correspond to percentiles of the statistical distributions for males or females. By this method one can test whether the design meets reasonable ergonomic requirements. Genetic algorithms are being used more and more to rapidly evaluate thousands of generations of systems successively modified by small perturbations to converge on better and better performance relative to some given constraints.

Education

To cope with the problems of humans (and, more generally, humanity) in relation to automation is to pose an ultimate challenge of educating users of technology and thereby educate the public at large. It is nothing short of the integration of science and humanities, the two cultures of C. P. Snow (1993). It is like bridging the ancient mind-body gap.

In this book I have tried in a very small way to do that: to give human factors engineers some hint of what automation promises for better or worse, and to give automation engineers something of the human perspective and an idea of why that perspective is important in system design. Human factors professionals must understand systems engineering concepts at a basic level, and Appendix A tries to help with this. It is just as important that systems engineers understand what human factors is about, especially consideration of the needs of the operator/user. I hope I have conveyed this in the body of the book.

It was proposed by one reviewer of this book that a brief summary of design caveats be included at the end. In a sense the whole book is caveats. However, here, at the conclusion, I will repeat a few general notions that I believe to be particularly salient for educating designers and users of future human-automation systems.

Experimentation and Modeling

To understand human-automation systems requires experimentation under scientific discipline. The laws of physics apply to the engineering of automation, but few quantitative laws are available for engineering the human's interaction with it. Systems analysis concepts, however, are available and apply to both. Both students and the general public need to understand the elements of these concepts.

Because of complexity and variability, pure determinism is seldom an option; one must resort to statistical inference. Technologists, as well as the educated public, should become more conversant with probabilistic thinking.

Errors are usually systemic, caused by the interactions of many latent factors that must be rooted out. Placing blame on those most immediate to any accident or incident seldom serves a useful purpose. What is more useful is to search out and correct the latent system deficiencies that provoke errors.

Human-in-the-loop simulation is an especially useful tool and has been employed widely for research and testing of hardware and software. Its advantages and limitations, particularly in terms of which fidelity simulator is appropriate for which problem, needs to be better appreciated.

Design
Human-automation system design is a synthesis of a relationship of the technology to human users. It is not building a thing. This is an important message for engineers to heed.

System design is a matter of trade-offs among many variables (physical, behavioral, economic), given the constraints posed by the management, the user, or whoever is posing the requirement. There are four components to this caveat:

a.   First, establish the hard constraints that are given, but don't assume constraints that are not really there. Task analysis is a good way to begin.
b.   Then think about the essential tradeoffs, the variables on the Pareto frontier, or the set of alternatives that are not dominated.
c.   Then think (or elicit, if practical) from the important affected parties their objective or utility functions.They will vary, and compromises will have to be made.
d.   Merge a, b, and c and satisfice.

If designers are not familiar with these decision concepts, they should be.

The designer should consider the decision criterion that best fits the situation − for example, what compromise to make between expected value (probability times consequences) and minimax. System designers are not accustomed to thinking in these terms.

In real human-automation systems "optimal" design is nonsense. So are notions such as "100% safety." One does one's best to lay out the variables, decision criteria, and compromises and then make a judgment.

Generic Caveats
Automation has wonderful benefits, but when applied carelessly, it can lead to complexity, unpredictability, and alienation in many forms.

The public needs to understand that automation is not all or none, and that the reality is almost always a human-automation interaction of some sort.

There has been a tendency to automate what is easiest and to leave the rest to the human. From one perspective this dignifies the human contribution. From another it may lead to a hodgepodge of partial automation, making the remaining human tasks less coherent and more complex than they need be, and resulting in overall degradation of system performance. Awareness of this tension is essential, as well as an appreciation that no simple solution exists.

Large-scale aggregates of humans and machines are particularly vulnerable to sometimes relatively simple terrorist acts. Much new thinking is required to reduce this vulnerability.

Systems engineers (planners, designers, managers) should assume responsibility for human-automation interaction. This may mean employing and working with human factors professionals, or it may mean the engineer becomes familiar with human engineering concepts. Preferably both.

Engineers by nature shy away from being advocates, and the advocacy for change is too often left to those with little technical training who really do not understand the technical issues involved. It is the duty of engineers as citizens to advocate for improvements that are within their domain of expertise.

A human's sense of productivity (fulfillment) may be orthogonal to machine (economic) productivity. The challenge is to make them correspond.

Appendix A

ELEMENTS OF THE MOST IMPORTANT NORMATIVE MODELS

Normative model means an abstract quantitative representation of how automation or human or a combination *should* work under specifiable assumptions. When stated mathematically, a normative model is simply a statement of what follows from the given assumptions. A *descriptive model* is a generalization of how something *actually works*, usually using mathematics to characterize a set of data. Often it is a best fit of a selected function (dependent variable function of an independent variable), such as a straight line.

The normative models described in Appendix A can be used to characterize human acquisition of information, decision, and control and to analyze machine acquisition of information, decision, and control, the component processes of automation.

I have selected for exposition an absolutely minimal set of models at a minimal level of detail, starting with probability and discrete decision models, without and with uncertainty, and ending with continuous control. If the reader understands these ideas, understanding human-automation analysis and design will come much easier.

1. Probability and Some Probabilistic Tests

1.1 Three Types of Probability
There are three kinds of probability:

Logical probability. This first type of probability is purely a mathematical one, defined as

$$p(x) = \text{limit as } N \rightarrow \infty \ (N_x / N), \tag{1}$$

where N_x is the total number of occurrences of an event of class x and N is the total number of events of all classes under consideration. This is a limit, and in real life one can never count to infinity, so logical probability is simply a useful fiction, like much of mathematics.

Empirical probability. This is the normalized frequency of events of class x:

$$f(x) = N_x/N, \tag{2}$$

where N_x is the empirical count of occurrences of events in class x and N is the total count of all occurrences.

Experimental data in a *sample* determine the empirical probability, from which one can only estimate logical probability of x in the *population* from which the sample is drawn.

Subjective probability. Subjective probability is a person's subjective estimate of probability given all the knowledge currently available. A rational person's subjective probability estimate is Bayesian (see the next section).

1.2 Bayes' Theorem and Updating

From basic combinatorics it is known that $p(A|B)p(B) = p(B|A)p(A) = p(A,B)$, where A and B can be any two events. The term $p(A|B)$ means probability of A given that B is known to be true, and $p(A,B)$ means the probability of A and B occurring jointly.

In its simplest form, Bayes' theorem (after a 19th-century English cleric whose hobby was math) is derived directly from the first two terms just given, namely:

$$p(A|B) = p(B|A)p(A)/p(B) \tag{3}$$

Substituting H for A and D for B — where H represents hypothesis, underlying truth, or cause and D represents data observed, apparent symptoms, or effect — results in

$$p(H|D) = p(D|H)p(H)/p(D). \tag{4}$$

This means that once some data (evidence) are available for a case of known H, one can refine any prior estimate of probability of H, $p(H)$, by multiplying by $p(D|H)$ and dividing by $p(D)$. The term $p(D|H)$ is what is known from a prior experience (i.e., where one knew H to be true and then observed D). The term $p(D)$ is the probability of those data being observed in general, independent of other circumstances.

Now suppose you observe D_1 and use Equation 4 to refine the prior knowledge $p(H)$ to get $p(H|D_1) = [p(D_1|H)p(H)/p(D_1)]$. Then you observe D_2. It is now possible to use the equation again to determine the new knowledge of H, $p(H|D_1D_2)$, by substituting $p(H|D_1)$ as the current prior knowledge of H,

$$p(H|D_1D_2) = p(D_2|H)[p(D_1|H)p(H)/p(D_1)]/p(D_2), \qquad (5)$$

where the term in square brackets is $p(H|D_1)$.

Suppose from the beginning you have two independent hypotheses, H_1 and H_2 , and went through the foregoing process in Equation 5. Then you could take a ratio of the two equations and would find that $p(D_1)$ and $p(D_2)$ would drop out because they are common to both equations, leaving

$$\frac{p(H_1|D_1D_2)}{p(H_2|D_1D_2)} = \frac{p(D_2|H_1)}{p(D_2|H_2)} \cdot \frac{p(D_1|H_1)}{p(D_1|H_2)} \cdot \frac{p(H_1)}{p(H_2)}. \qquad (6)$$

The first term is called the *posterior odds ratio*. The last term is the *prior odds ratio*. The two terms in between are called *likelihood ratios* for D_2 and D_1 , respectively. This process can be extended to any number of data (they don't have to be independent of one another) to find a posterior odds ratio. In general, the posterior odds ratio is the product of likelihood ratios for all salient data times the prior odds ratio.

This formulation is independent of the order in which the data (evidence) are observed. It assumes the process statistics are stationary (don't change with time). Old data count just as much as new data. If desirable, given that the world is not always stationary, a Bayesian updating process can be adjusted to discount data according to how old it is.

1.3 Optimal Sampling

How often should a machine or a person sample a continuous signal to be able to capture all the information in it (i.e., to reproduce it accurately)? The answer can be demonstrated by a simple example.

Assume an arbitrary time function of duration T. Suppose such time segments are strung together, forming an infinitely long periodic function $f(t)$. This function can be represented by a Fourier sine-cosine series:

$$f(t) = \sum_{n=0 \text{ to } \infty} A_n\cos(n\omega t) + \sum_{(n=0 \text{ to } \infty)} B_n\sin(n\omega t), \qquad (7)$$

where $\omega = 1/T$. Because it is a repeated function, each T-long segment is represented by $f(t)$. Now further assume the signal is band limited to ω_{max} (there are no frequency components higher than this). That means that $f(t)$ has $(\omega_{max}/\omega) = (\omega_{max}T) = \cos$ terms and the same number of sin terms, or $(2\omega_{max}T)$ terms per segment $= (2\omega_{max})$ terms per unit time.

This generalizes to the idea that for any band-limited signal, all the information can be obtained (the signal can be perfectly reproduced) by accurate observations at a sampling rate of $(2\omega_{max})$.

Senders et al. (1964) demonstrated through eye movement studies that human participants divide their sampling among multiple instruments of different bandwidths according to the bandwidth of the signal they are sampling, slightly oversampling slower signals and undersampling faster ones. Carbonell (1966) and Sheridan (1970) proposed more complex models for how often one should or does sample dynamic functions, taking into account value considerations that will be discussed in a later section. Other models were reviewed by Moray (1986).

1.4 Expected Value of the Cost Incurred from an Erroneous Response

The expected value of cost incurred when a person or a machine responds in an erroneous or inappropriate way is often of great significance in human-automation systems. It is the product of four terms,

$$p(O)\ p(E|O)\ p(NR|E,O)\ C, \tag{8}$$

where $p(O)$ is the probability of *being provided the opportunity* to respond (voluntary or involuntary exposure, knowledge of the opportunity, time to respond, etc.); $p(E|O)$ is the probability of *making an error, given the opportunity*; $p(NR|E,O)$ is the probability of *not recovering in time*, given that the opportunity occurred and the error was committed; and C is the *cost* — when there was opportunity or exposure, the error occurred, and there was no recovery.

1.5 Null Hypothesis Testing and Its Pitfalls

Null hypothesis testing is commonly regarded as the core of practical statistics. The general procedure and its logic go as follows:

1. From experimental data determine some aggregate measure (called a *statistic*).
2. Normalize the statistic to remove dimensions and find its corresponding fractile (or percentile) on an appropriate cumulative probability function. For example
   a. In testing for significance of difference between means of samples that have been subjected to different experimental treatments, or between an experimental mean and a known population mean, one normalizes by an estimate of the population standard deviation (of a distribution of that difference between means, usually obtained for the experimental data) and determines at what fractile the result falls on a near-Gaussian or Student's t distribution.
   b. In testing whether the frequency of occurrences of data in a number of experimental (mutually exclusive and collectively exhaustive) categories significantly deviates from expected frequencies, one normalizes by the degrees of freedom and determines the fractile on a chi-square distribution.

c. In testing whether the variance of sample means for different experimental treatments is significantly larger than what would result if the treatments had no effect, one normalizes by a pooled variance estimate among participants within treatments and checks the result against fractiles of Fisher's F (or variance ratio) distribution. The details of obtaining the normalizing factor for a given test can be found in any elementary statistics book.

3. The null hypothesis is assumed — that the obtained statistic occurred because of chance variation, there being no systematic effect of treatments. If the statistic is too rare — that is, if it is out beyond the predetermined criterion on either of the tails of the assumed distribution of comparable statistics (such as x_1 or x_r in Figure A1), one rejects the null hypothesis. Otherwise one accepts. (Note: Sometimes only one tail is appropriate.)

In rejecting the null hypothesis, one always risks a certain probability of being wrong $(1 - p_r)$, for the right tail and/or p_l if a left tail is appropriate. This is the risk that there really was no treatment effect and the extreme statistic did actually occur by chance. This risk probability of a rejected but true null hypothesis is called a *Type I error*. One typically sets a Type I error as a criterion of statistical significance. The farther out on the tail (or tails) one sets the criterion of rejection, the tighter (more conservative) the requirement for statistical significance.

A common pitfall is not realizing that Type I errors will occur $\{(1 - p_r) + p_l\}$ percentage of the time in null hypothesis tests. This error is essentially what is set by the significance level criterion.

Another, more subtle, pitfall is the *Type II error*, which occurs when the null hypothesis is accepted (the statistic is not seen as rare enough to pass the significance test), though in fact the treatments really did have an effect. There is no simple remedy for this type of error.

A third pitfall is that with a large enough sample size, even though the treatment differences have very little effect, one can almost always prove

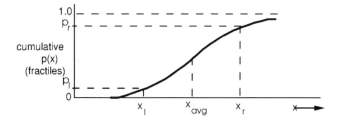

Figure A1. Example of cumulative probability distribution (integral of the density function) with left- and right-tail criteria shown — for example, 5th and 95th percentiles. The curve in this case is symmetric and Gaussian in shape.

significance. Considering differences between treatment means, for example, the normalization factor gets smaller in proportion to the inverse of the square root of the sample size, so the larger the sample size, the larger the normalized statistic, other things being equal. For this reason some experimentalists are skeptical about the meaning of *significance* in null hypothesis tests with large samples.

2. Information

2.1 Information as the Quality of Measurement

Information about some object or event is gained by measurement. The measures, whether by human or machine, can have very different qualities. Another way of saying this is that the numbers (or symbols) by which the measured object or event is scaled can have very different meanings with respect to one another. The classical taxonomy of such measurement (scaling) has four categories: *nominal scales*, *ordinal scales*, *interval scales*, and *ratio scales*. The distinctions among these scales are made by the permissible transformations on the numbers.

A *nominal scale* is a naming process. The object or event is assigned a unique number. For a given set of such operations on different objects or events, any transformation (new set of number assignments) is permissible as long as every object or event has its own unique number. Examples are Social Security numbers and numbers on football players' shirts.

An *ordinal scale* assigns numbers to order the objects or events. A greater number means a greater value of some attribute. Any transformation is permissible as long as monotonicity and the same ordering are preserved. Examples are quality ratings for hotels or films and Brinell hardness of surfaces.

An *interval scale* assigns numbers to the objects or events in such a way that equality of numerical intervals is preserved. Any transformation is permissible that maintains the equality of intervals between any set of objects or events. For example, if A and B are 2 feet apart and B and C are 2 feet apart, one can transform to inches or meters and the $A - B$ and $B - C$ intervals will still be the same number (as each other) apart. Both intervals will be 2 feet, 24 inches or 0.6095 meters. Note that an arbitrary number can be added to each of A, B, and C and the intervals between them would not change. Temperature measured in degrees Fahrenheit would be another example of an interval scale.

A *ratio scale* assigns numbers to objects or events relative to some absolute zero reference, which means that under any transformation, the ratios are preserved. Note that if $A = 4$ and $B = 2$, $A = 2B$. Transformation to a scale with half-size units would make $A = 8$ and $B = 4$, so A is still

$2B$. Note that in this case an arbitrary number cannot be added to each of A and B, for the ratio property would not then be preserved. Temperature measured in degrees Kelvin would be an example of a ratio scale. On the Kelvin scale the zero has a meaning, and a number twice as large as another number is truly twice as hot. On the Fahrenheit scale that would not make sense — a number twice as large as another number does not mean twice as hot in any meaningful way. Note further that each successive scale (in the order just given) retains the properties of the one preceding it.

2.2 Information as Correlation: The Brunswik Lens Model

Although correlation between two sets of items does not necessarily mean causality, it is strongly suggestive, and one tends to make the reasonable assumption of strong connotation. For a person to make judgments about some environmental state or property, a cascade of three transformations must occur. First, there must be some physically observable cues or evidence that lend more credence to one hypothesis than another. Second, the human must access those cues (see, hear, or touch them) and weight their relevance to the discrimination to be made. Third, the human must be able to make use of the accessed and weighted cues to render a specific judgment. Figure A2 conveys the idea using the metaphor of an optical lens, the concept being a contribution of the psychologist Brunswik (1956) with quantitative refinement by Hursch, Hammond, and Hursch (1964).

The idea is that physical reality (at left in Figure A2) is seen through a lens, and the effects (rays, heavy lines at left) of the present physical state or property of interest generate a set of cues (small dark circles on the left

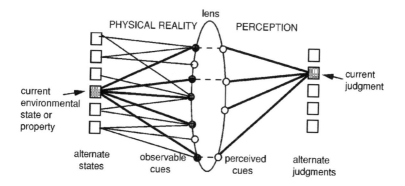

Figure A2. Interpretation of Brunswik lens model.

surface of the lens) that are potentially observable. At other times other states also generate cues, some of them the same, some different from the present set (light circles on the left side). This is the first transformation.

All cues from the present left-hand set are not necessarily perceived or properly weighted as relevant by the human (circles on the right surface of the lens), so that some cues are mistakenly perceived or weighted (circles not connected by dashed lines). In any case the human ends up focusing on a set of perceived cues on the right side of the lens. This is the second transformation.

The final transformation is using the perceived cues to make a judgment. (The correlation between states on the left and judgments on the right is the overall performance of the information acquisition plus judgment.)

Mathematically, the correlation between judgment and state is

$$R = R_1 G R_2 + C(1 - R_1)^{0.5}(1 - R_2)^{0.5}, \tag{9}$$

where R_1 is the state-to-observables correlation at left, R_2 is the perceived cues-to-judgment correlation at right, and G is the correlation between the cue weighting generated by the environment and the cue weighting corresponding to the judgment (Cooksey, 1996). The first term is the linear additive correlation from state to judgment. The second term is a correction factor, where C is found as the positive or negative correlation between the residuals of the human and environment models.

As another correlation, C is in the range $(-1, +1)$. Usually C is found to be essentially zero, indicating that linear-additive models are sufficient to describe judgment. However, there are cases in the literature when C is significant (say, 0.4 or 0.5), indicating that humans can sometimes make use of more complex (e.g., curvilinear) patterns in the relationship between the environment cues and a state.

The advantage of this model is that it combines characterization of the environment with characterization of the human (Bisantz et al., 2000).

Measures other than correlation may also be utilized. For example, given the conditional probabilities of various cues for various states, Bayesian updating (see the previous section) can be employed to assess the relative probabilities of alternative states. Given joint probabilities for input-output combinations for each of the three cascaded stages, Shannon (1949) information theory (see the next section) can be used to analyze not only transmitted information but also information loss or equivocation (e.g., the shaded cue at left without counterpart cue at right) and noise (e.g., the cue at right without counterpart shaded cue at left).

2.3 Information as Uncertainty (Shannon Information)

Information, as defined by Shannon (1949), can be thought of as the degree of uncertainty about an event before it happens. If, after it happens, the event is known with certainty, that same measure is a measure of surprise, or uncertainty reduced, or *information transmitted.* If one of two equally likely events occurs (a one-bit reduction of uncertainty), there is less surprise than if 1 of 16 equally likely events occurs. (The latter is a four-bit reduction of uncertainty, where number of bits $= \log_2 N$, in this case $\log_2 16 = 4$.)

Automatic systems need information to work. Within certain limits, an automatic system can adjust to the surprise, assuming it can be measured, and provide a response that either maintains some variable constant or tracks the changes that occur. If the surprise cannot be measured, then that input is regarded as noise and the automation can still be configured to be unaffected by (reject) the noise. In a later section I will explain how.

The accepted Shannon definition of *average information H* of a *message set* is

$$H = \sum_i p(x_i)\log_2[1/p(x_i)] = -\sum_i p(x_i)\log_2 p(x_i). \qquad (10)$$

(Note that this H for information has a different meaning than the H for hypothesis in Bayesian manipulations.) The Shannon measure indicates the average uncertainty of the observer (recipient) before receipt of the message (revealing which x_i is the correct one), given a set of possible messages and their probabilities, $p(x_i)$. If, after the message is sent, all messages in the set have probability zero except for one that has probability one, then the uncertainty is zero. Thus the message set is defined by its uncertainty-reducing capability.

When all event states have equal probability, then $\log_2[1/p(x_i)] = \log_2 N$. This term corresponds to the number of steps in a split-half procedure for narrowing down from the initial N equally probable states to the final certainty of which state is the true one. The H defined here for (in general) unequally probable states can also be called H_{avg} because it is a weighted average over log transformations on component probabilities.

Shannon information theory also includes concepts of information transmission, noise, and equivocation. Starting with information on sets defined as follows,

$$\textit{input information:} \quad H(x) = \sum_i p(x_i)\log_2[1/p(x_i)] \qquad (11)$$

$$\text{output information:} \quad H(y) = \sum_j p(y_j)\log_2[1/p(y_j)] \qquad (12)$$

$$\text{joint information:} \quad H(x,y) = \sum_{ij} p(x_i,y_j)\log_2[1/p(x_i,y_j)], \qquad (13)$$

Shannon defined *transmitted information* as

$$H(x{:}y) = \sum_{ij} p(x_i,y_j)\log_2[p(x_i|y_j)/p(x_i)], \qquad (14)$$

where $p(x_i,y_j)$ means probability of x_i and y_j occurring together and $p(x_i|y_j)$ means probability of x_i given that y_j is true. The vertical bar always means "given" and will be used for other equations to follow. The expression given in Equation 14 for information transmission amounts to the difference between uncertainty before a message is sent and uncertainty after — a reduction in uncertainty and hence transmitted information. (Because of the logarithm, the difference becomes a ratio of the terms in square brackets.) Note that after a message is sent, the uncertainty is not necessarily zero. Because $p(x_i|y_j) = p(x_i,y_j)/p(y_j)$,

$$H(x{:}y) = \sum_{ij} p(x_i,y_j)\log_2[p(x_i,y_j)/p(x_i)p(y_j)], \qquad (15)$$

so that

$$H(x{:}y) = H(x) + H(y) - H(x,y). \qquad (16)$$

Another useful definition is

$$\text{noise:} \quad H(y|x) = \sum_{ij} p(x_i,y_j)\log_2[1/p(y_j|x_i)]. \qquad (17)$$

This is a measure of uncertainty about the output after the input is given (fixed). In other words, noise is the spurious uncertainty or variety in the output that crept into the process somewhere in the middle. Finally, there is

$$\text{equivocation:} \quad H(x|y) = \sum_{ij} p(x_i,y_j)\log_2[1/p(x_i|y_j)]. \qquad (18)$$

This is uncertainty about the input after the output is given. In other words, it is input variety that has no effect on output variety, variety that is lost in the process.

An important general property of Shannon information is that transmitted = input − equivocation = output − noise,

$$H(x:y) = H(x) - H(x|y) = H(y) - H(y|x), \tag{19}$$

which adapts to the summary diagram in Figure A3. This relation can be derived algebraically from the definitions just given.

Readers interested in applications of these concepts to human-machine systems are referred to Sheridan and Ferrell (1974).

2.4 Information Value

Information value, based on the fundamental decision problem explicated by Raiffa and Schlaifer (1961) and later elaborated by Howard (1966) and others, refers to the difference in what one can gain by action taken knowing the state of the event, relative to what one can gain by action taken without such knowledge. Let $V(u_j|x_i)$ be the gain or reward for taking action u_j when an event x is in state i, and let V be money, time, energy, or any measure of value, including the economist's combined measure, *utility* (which is discussed in the next section). If x_i is known exactly, then a rational decision maker adjusts u_j (selects j) to maximize V for each occurrence of x_i , in each instance yielding $\max_j V(u_j|x_i)$. In this case the average reward over a set of x_i is

$$V_{\text{avg}} = \sum_i p(x_i)\{\max_j V(u_j|x_i)\}. \tag{20}$$

If x_i is known only as a probability density, $p(x_i)$, then the best a rational decision maker can do is to adjust u_j once, to be the best in consideration of the whole density function $p(x_i)$. In this case the average reward over a set of x_i is

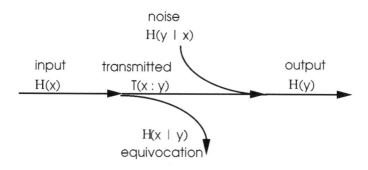

Figure A3. Information relationships.

$$V'_{avg} = \max_j \left\{ \sum_i p(x_i)V(u_j|x_i) \right\}. \tag{21}$$

Information value, then, is the difference between the gain in taking the best action given each specific x_i as it occurs, and the gain in taking the best action in ignorance of each specific x_i — that is, knowing only $p(x_i)$. This difference is

$$V^*_{avg} = V_{avg} - V'_{avg}. \tag{22}$$

The two concepts, information and information value, characterize different aspects of information: seeking and using (Sheridan, 1995). The Shannon measure of information (uncertainty reduction) is a measure of the processing effort required to discover the truth from an initially uncertain state. It can be applied to specification of x_i by a machine or by cognitive effort — for instance, in the former case by providing the information in a display, in the latter case by retrieving information from memory or by finding information by visual search. It says nothing about what one can gain by having the information — that is, by taking action based on the information.

Just as clearly, the idea of information value as stated regards only what can be gained by taking action based on the information as compared with taking action without it. It can also include the cost of action. However, it says nothing about the processing effort required to reduce uncertainty. Differences between *information* in the Shannon sense and *information value* are elaborated by Sheridan (1995).

A further consideration is the cost (in the same dimensions as V) to get the information. In application, the cost of information subtracts from original information value to yield a net information value. To achieve break even, one might argue that the information value is what one should be willing to pay a clairvoyant (or the researcher, provided he or she comes up with the right answer).

3. Utility Function (Objective Function)

The V terms used in the last section might have been regarded as having dimensions of dollars; however, not all objects and events come with a prespecified dollar value. There is somehow an intrinsic value, a subjective judgment of *relative worth*, that is different for different people and even different for the same person under different circumstances (e.g., when you really need a coin for the telephone vs. when you don't). Furthermore, even considering dollars, a 50% chance of $2X$ dollars is not valued the same as 100% chance of X dollars, even though they have the same expected value. The same would be true for ice cream cones, the size of one's house, walks in the park, or anything else with a quantifiable attribute.

3.1 Von Neumann Axiom of Utility and How to Use It

The problem stated in the preceding paragraph plagued thinkers about relative worth and fairness and economic welfare for 100 years, until Von Neumann and Morganstern (1944) defined *utility* by an axiom and an experimental implementation of it. In its simplest form, and for a single attribute (assume dollars for now), the axiom specifies the utility U of an object or event C occurring with certainty, in terms of a lottery of two other objects or events and their known utilities:

$$U(C) = pU(A) + (1-p)U(B), \tag{23}$$

where U is measured in some abstract dimension; call it *utiles*. As the reader will see, this equation has its meaning only in relative terms, where p and $(1-p)$ are probabilities of mutually exclusive and collectively exhaustive events A and B.

The simplest way to use this idea is to let A and B be extremes of the range of the attribute on which you wish to scale utility, say $A = \$100,000$ dollars and $B = \$0$. Define A to have a utility of one and B to have a utility of zero. If you let $p = .5$, then according to the definition of utility, $U(C) = .5$. In terms of Von Neumann's axiom, that means you would be indifferent between having C for sure and a 50-50 lottery of $0 and $100,000, as represented by the upper part of Figure A4. The question is,

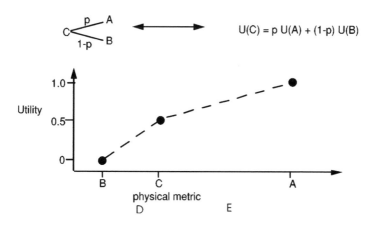

Figure A4. The definition and experimental elicitation of a person's judgment of utility.

what is the corresponding *C*? In implementing an indifference judgment experiment, most people would select a point along the physical attribute axis in Figure A4 significantly less than $50,000, as shown. One would not want to risk gaining nothing. From this one can note that the utility curve is unlikely to be a straight line.

Such a concave downward curve means the decision maker is *risk averse*, opting for something less than expected value in order not to risk gaining nothing. Risk aversion is why people purchase insurance, paying the insurance company to avoid such a risk. The insurance company, by insuring many clients, can afford to play on an expected value basis. The individual risk taker cannot afford that luxury.

This same procedure can be repeated to determine other points on the utility curve. To determine the location of a point *D* between *B* and *C*, *C* and *U(C)* are taken to represent the maximum of the range on each axis, respectively. To determine the location of a point *E* between *C* and *A*, *C* and *U(C)* are taken to represent the minimum of the range on their axes. Theoretically one can continue in this manner, but because of experimental error, only a few data points are usually taken between zero and one for any attribute.

It should be noted that this procedure assumes smooth monotonicity between the anchor points initially assumed (*B* and *A* in this example). If for some attribute there is a known maximum utility in between the anchor points, say at *F*, two utility functions should be determined: one between *B* and *F*, the second between *F* and *A*. Examples of such attributes might be the number of calories in a meal, or one's desired weight, where an intermediate value is preferred to either extreme.

3.2 Multiattribute Utility

Sometimes two or more attributes are combined to determine a scalar *multiattribute utility function*, also called a *trade-off function* or *objective function*. There are procedures to determine the utility of points within an attribute space of two or more dimensions. The simplest procedure might be a simple linear weighting function of the two attributes, and although that is often used as a first approximation, it is no more valid than a linear utility function for a single attribute.

A more rigorous procedure goes as follows, assuming for purposes of explanation a two-attribute case: First, the utilities of the two maximum single-attribute points are determined by using the same procedure as before but letting *p* be the variable judged in each case. Then, with all the corner points known, utility functions for all the edges of the attribute are determined as before. Finally, to determine the utility of points within the rectangle, a procedure is invoked that assumes that the *shape* of an interpolation utility curve for intermediate cross-sections is the same as the shape of the single-attribute utility curves, except in this case the end points of the cross-sectional interpolation curve are determined from known opposite points along the opposing edges. This is called the *quasi-separability* assumption (Keeney & Raiffa, 1976).

It should be clear that utility need not always be positive; it can just as easily be negative. Next, an example will be used in which attributes are two "costs" that are not reconcilable except by means of subjective utility judgment.

4. Discrete Decisions

4.1 Decision under Certainty: Selection along a Pareto Frontier

Decision under certainty means everything is known; there are no probabilities to contend with, only the selection among available alternatives. Consider Figure A5 a decision space, in this case with points (letters) representing alternative used trucks a person is considering purchasing, any of which might do the job.

To make the example simple, only two attributes are shown: maintenance cost and initial cost. Assume that only a finite number of trucks are available, so the decision maker must decide among those combinations shown.

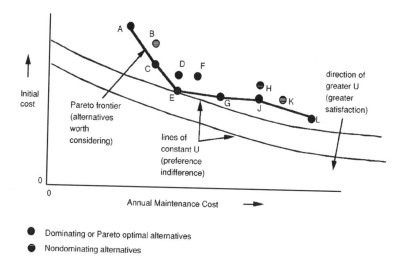

Figure A5. Pareto frontier and utility curve intersection determines optimal choice.

Notice that some choices *dominate* other choices: For the same maintenance cost, Truck C has a lower initial cost than Truck B. A rational decision maker would always prefer Truck C over Truck B. Truck E similarly dominates Truck D. Truck J dominates Truck K based on maintenance cost and Truck H based on initial cost. Assuming Truck J has very slightly lower initial cost than G, Trucks A, C, E, G, J, and L are not dominated. Together they constitute a set of nondominated alternatives that trade off initial cost and maintenance cost. No purpose is served by considering the other alternatives.

The decision maker can get lower initial cost only at higher maintenance cost. The points on the heavy line are said to form a *Pareto frontier* (named after an Italian economist, Vilfredo Pareto). The rational decision maker rejects whatever is up and to the right of this line and wants what is down and to the left. Unfortunately, there is no truck with low initial cost *and* low maintenance. The Pareto frontier defines the best that can be had, the set of alternatives from which to choose, but still leaves the decision maker having to decide among the points on this line.

How to decide? The rational choice is to select the alternative with the best utility. The two light and smooth-curved lines are hypothetical constant utility curves, determined, for example, by the Von Neumann procedure. Each is a line of utility indifference. For a set of such lines, "better" is down and to the left because the attributes are costs. Therefore, the best choice is the point of intersection of the Pareto curve with the indifference curve of best utility, in this case Truck E.

One need not draw the Pareto curve explicitly, because one could go directly to the alternative with best utility. When there is a large number of alternatives or a continuous space, however, it is helpful for the decision maker to consider the Pareto frontier.

Note that there are two kinds of trade-off curves in decisions under certainty: (a) the Pareto frontier, which rejects dominated alternatives and represents a trade between alternatives based only on the direction of better or worse but has nothing to do with the subjective utility trade between attributes; and (b) the set of utility indifference curves, which represent the subjective trade between attributes but say nothing about available alternatives.

4.2 Crisp Rule-Based Decision

Decision making can be done according to a set of prearranged rules. The reasons to do this are to provide order, anticipate circumstances, and not have all decision making be ad hoc. It is why governments have laws. Powerful computers allow the construction of information systems that make use of stored rules and respond to queries from humans or trigger information processing and display based on artificial sensors. These are variously called *expert systems, decision support systems,* and *knowledge-based systems.* In such rule-based systems the computer looks at its inputs, searches for applicable stored rules that might apply to those inputs, and uses some synthesis of those rules to determine the output.

Rules can be *crisp* or *fuzzy,* depending on the logic used. Precise quantification and Boolean logic imply crisp sets, wherein a number or symbol is identified with a well-defined set of objects or events and is explicitly not identified with other objects or events. Any one object or event either is in the set corresponding to a given symbol or it is not; one can say its membership M in the set is either 0 or 1. Crisp rules combine logical ANDs and ORs, such as "If (a AND b) OR c, do d, ELSE [meaning if none of the aforementioned conditions is true] do e."

Note, by the way, that if the rule were differently parsed, such as "If a AND (b OR c)," its meaning would be different and that "If a AND b OR c" is actually ambiguous.

If multiple rules might apply but conflict in their outputs, then there must be some precedence rule, weighting of actions, or other means of deciding what to do.

Computer programs having such rules may be thought of as automation, provided they are connected to the world such that their electronic decisions actually *do* something.

4.3 Fuzzy Rule-Based Decision

Fuzzy rules are different from crisp rules because fuzzy sets are fundamentally different from crisp sets. Fuzzy mathematics was introduced by Zadeh (1965). It has been extensively developed in Japan and other nations — perhaps more than in the United States, where it is controversial. For some people the term *fuzzy* seems a contradiction to rationality (which it is not!).

The membership M of objects or events of a fuzzy set can be 0, 1, or anything in between. The meanings of words (the symbols of natural language) are fuzzy in the sense that a word can apply very well (clearly, obviously) to some objects or events, can clearly exclude other objects or events, and can apply somewhat (more or less, partially) to yet other objects and events.

Figure A6 gives a simple example of plausible meanings, in terms of membership functions M (relative truth, applicability, etc.) for each of two physical variables x and y relating to landing a private aircraft at a nearby field rather than going on to the planned destination. The first physical variable x is gradations of engine roughness (e.g., as measured by an accelerometer), and the second physical variable y is gradations of remaining fuel (e.g., as measured in gallons). Each of the four curves for engine roughness and the four for remaining fuel specifies membership in a fuzzy set (each labeled by a corresponding verbal term) relative to the associated quantitative and objectively measurable variable.

Then, any given quantitative value of a single physical variable maps to a fuzzy vector **M** of different (scalar) memberships M corresponding to each of several different fuzzy terms. In this example, the engine roughness indicated by the mark "x" on the abscissa of Figure A6 has membership vector $\mathbf{M} = .8$ for "somewhat rough," .4 for "OK," and .2 for "very rough." The fuzzy vector for the fuel mark "y" maps to $\mathbf{M} = .9$ for "little fuel," .6 for "critical," and .2 for "normal." For the physical state vector (x,y) for engine roughness and fuel combined, we get a combined vector.

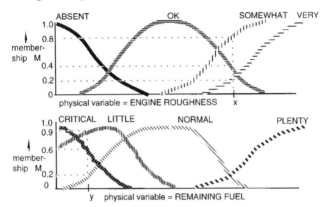

Figure A6. Fuzzy membership functions and procedure to evaluate landing rules.

Evaluate membership for given case, using OR=max, AND=min:

(1) "If engine is somewhat rough or very rough, and there is little remaining fuel.........or, if fuel is critical, then land the aircraft."
(2) "However, if engine roughness is OK or absent, and fuel is normal or plenty, continue flight." (Note that these two rules do not cover all situations.)

For first rule: If {[(somewhat) OR (very)] AND [little]} OR {critical}, then land. Similar for second rule.

(1) max [(somewhat), (very) roughness] = max [0.8, 0.2] = 0.8,
         min [(0.8), (little fuel)] = min [0.8, 0,9] = 0.8
(2) max [(OK), (absent) roughness] = max [0.4, 0] = 0.4
         max [(normal), (plenty)] = max [0.2, 0] = 0.2
         min [0.4, 0.2] = 0.2 for continue flight

Take action with greatest membership (land airplane), given that all available rules were evaluated for case at hand. Or, if evaluation of available rules for this case results in no membership greater than some criterion, invoke other procedure.

As shown in the following paragraphs, it follows that any logical "if-then" statement made up of fuzzy symbols (e.g., "If the engine is somewhat rough or very rough and fuel is getting low, or if fuel is short, then land") yields a net membership scalar (or relative applicability) for that particular physical situation. It is common to assume that the conjunction OR (logical union) means the *maximum* of the two associated terms, and the conjunction AND (logical intersection) means the *minimum* of the two terms. Then we can evaluate the combined vector and determine the net *M* scalar for the given rule by parsing:

{[(somewhat rough) OR (very rough)] AND [(getting low on fuel)]}

OR {(fuel critical)} = land.

This statement, for specified values of roughness and fuel, therefore has relative membership

$$M = \max\{\min[\max(.8),(.2)],[(.9)]\},\{(0.6)\} = 0.8. \tag{24}$$

The figure shows the separate steps involved.

Other rules may recommend different actions — for example, "If *A* is the case, then do U_1 ; if *B* is the case, then do U_2 ." One procedure for determining final action in a given situation is to let the rule or combination of rules that have relatively greatest applicability dominate. In other words, the action with the greatest net membership *M* should be taken, as is done in the text under Figure A6.

Another procedure, applicable when the candidate actions are different degrees of the same physical variable, is to weight each action according to its membership so that the final action is a compromise (weighted sum).

4.4 Decision under Uncertainty

Until now I have been talking about decision under certainty (i.e., the consequences that follow from any decision are known with a probability of one). The only question is what decision is best. Even in the case of fuzzy rules, the memberships are all known.

When the states of the world and consequences of each decision are known only by probabilities that are less than one (the decision must be made without knowing for sure what will happen), this is *decision under uncertainty*. Under uncertainty any decision option has attached to it two or more contingencies, each of which has a probability. This is represented in Figure A7, sometimes called a *payoff matrix*. In this example three decision options are shown (A, B, and C) and two independent contingencies (not under the decision maker's control): *X*, which is known to occur with probability .3, and *Y*, which is known to occur with probability .7. The numbers in the cells are the payoffs (positive utilities this time) for each combination of contingency and decision. The decision maker wishes to maximize payoff.

Normally, especially if one expects to make repetitive decisions in the same environment of contingencies and payoffs, one selects the decision that has the highest expected value (first column to the right of the box in Figure A7). This would be Decision C, for which the expected value is .7(6) + .3(3) = 5.1.

4.5 Decision Criterion and Optimization

Expected value is not the only *decision criterion;* others can make sense, depending on the decision maker's situation. For example, suppose the decision maker finds great satisfaction in gambling. He or she would select Decision B, which, if it wins, would produce the maximum possible gain (see the second column at the right in Figure A7). Or consider a third decision maker, a man who is a pauper. He is not so concerned about winning the most possible; he is concerned about the downside risk, ending up in the worst case. He is risk averse. In that case he would use a *minimax* criterion to maximize the minimum possible gain. In other words, he would determine the minimum payoff for each decision and select that decision with the greatest of these minima — to be sure to earn at least that much. This obviously would be Decision A (see the third column in Figure A7).

Thus it is clear that with precisely the same decision situation, when alternatives, probabilities, and payoffs are all given, the optimum decision can be quite different as a function of the decision criterion.

Optimization is a trivial exercise if explicit utilities, explicit probabilities, or other constraints are given (e.g., if the decision maker must earn at

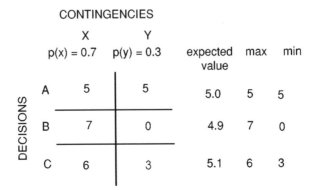

Figure A7. *Payoff matrix for decisions under probabilistic contingencies.*

least Z or else not play, or if one alternative is illegal). One merely maximizes the utility for the possible decision alternatives subject to the various probabilistic and/or deterministic constraints, including the decision criterion. That is precisely what is done in linear programming, in many other forms of operations research (dynamic programming, game theory, etc.), and in optimal control. The constraints in this case are usually determined by the laws of physics and/or economics. Quite likely the constraints are themselves relations among the salient variables: Newton's law $F=MA$, Ohm's law $E=IR$, and so on. What engineering analysis is all about is specifying these constraining relations. However, the problem is often less a matter of specifying the constraints than of discovering what the salient subjective utilities are.

4.6 An Example of Applying Statistical Decision Theory: Whether to Automate

If the question is simply whether or not to automate, a straightforward design procedure is available by applying standard statistical decision theory to the failure response task (Sheridan & Parasuraman, 2000). Let the true states of the world be represented by F for *failure* or *abnormal* and N for *no failure* or *normal*. Further, let $\underline{F}$ or $\underline{N}$ mean *decision* or *response* (appropriate to F or N, respectively) in the failure response task. One wishes to maximize the expected value (EV) of human or automation performance of the task. If the benefits of correct responses and the costs of incorrect responses are represented by B and C, respectively, then

$$EV=p(F)p(\underline{F}|F)B(\underline{F}|F)+p(N)p(\underline{N}|N)B(\underline{N}|N)$$
$$-p(F)p(\underline{N}|F)C(\underline{N}|F)-p(N)p(\underline{F}|N)C(\underline{F}|N). \qquad (25)$$

In words, there are four components that add to constitute the expected value. Each is composed of three factors: (a) a priori probability of actual state, (b) conditional probability of correct or incorrect decision/response, and (c) benefit or cost of correct or incorrect decision/response. Respectively, the four independent terms of Equation 25 are commonly known as the expected return for true positive, true negative, false negative (or miss), and false positive (or false alarm) responses.

In each of the four terms only the second factor characterizes the human or automatic decision maker. The first and third factors characterize the given situation.

Because $p(\underline{F}|F)=1-p(\underline{N}|F)$ and $p(\underline{N}|N)=1-p(\underline{F}|N)$, Equation 25 can be simplified to

$$EV = p(F)B(\underline{F}|F) - p(F)p(\underline{N}|F)[B(\underline{F}|F) + C(\underline{N}|F)]$$
$$+ p(N)B(\underline{N}|N) - p(N)p(\underline{F}|N)[B(\underline{N}|N) + C(\underline{F}|N)] \qquad (26)$$

The first and third terms in Equation 26 are the same whether the system is automated or not. The second and fourth terms represent lost gains and/or real costs that depend on failure response probabilities that are likely to be different for automation as compared with human control. Hence it can be asserted that automation should be used if the expected costs for automation are less than those for human control, or if

$$p(F)p(\underline{N}|F)_A[B(\underline{F}|F) + C(\underline{N}|F)] + p(N)p(\underline{F}|N)_A[B(\underline{N}|N) + C(\underline{F}|N)]$$
$$< p(F)p(\underline{N}|F)_H[B(\underline{F}|F) + C(\underline{N}|F) + p(N)p(\underline{F}|N)_H[B(\underline{N}|N)$$
$$+ C(\underline{F}|N), \qquad (27)$$

where subscripts A and H represent failure response probabilities for automation and human control, respectively.

Rearranging terms and substituting $p(N) = 1 - p(F)$, the rule becomes

Automate if $[p(\underline{N}|F)_A - p(\underline{N}|F)_H] < K[p(\underline{F}|N)_H - p(\underline{F}|N)_A]$, $\qquad (28)$

where

$$K = \frac{[1 - p(F)][B(\underline{N}|N) + C(\underline{F}|N)]}{p(F)[B(\underline{F}|F) + C(\underline{N}|F)]}. \qquad (29)$$

The left- and right-side bracketed terms in Equation 28 can be either positive or negative, because these represent *differences* in incorrect failure response probabilities (false negatives and false positives) between human and automation control, each of which may be greater or lesser than the other. The multiplicative term K, which is derived in Equation 29, is always positive because the benefits B and costs C are defined as positive numbers and because $1 - p(F) > 0$.

Equation 28 can be stated in words as follows: Automate if the difference in false negative probabilities between automation and human control is less than the weighted difference in false positive probabilities between human and automation control, where the weighting term K is dependent on the a priori probability of a failure state and the benefits and costs.

The term K is equivalent to the term known as β in signal detection theory (see the next section; Green & Swets, 1966).

Limiting cases for $p(F)$. It is useful to consider the limiting cases of a very high a priori probability of a failure state and very low failure probability. From Equation 28, one can readily deduce, for $p(F) \rightarrow 1$, automate if $p(\underline{N}|F)_A < p(\underline{N}|F)_H$. $\qquad$ (30)

In this case only false negatives (misses) are of concern. Equation 30 suggests that for systems with high a priori probabilities of failure for particular events, automation should be pursued whenever human detection of the event is likely to be poor or not timely. In such cases, automation is mandated irrespective of the false positive rate of automation.

For $p(F) \rightarrow 0$, automate if $p(\underline{F}|N)_A < p(\underline{F}|N)_H$. $\qquad$ (31)

In this case only false positives (false alarms) are of concern. Equation 31 indicates that for systems with low a priori probabilities of failure for particular events, the false positive rate of automation is critical to the decision to automate. Unfortunately, many automated warning systems, such as collision warning systems for aircraft and automobiles, often have high false alarm rates, thereby leading to disuse or disablement by human operators (Parasuraman & Riley, 1997). Parasuraman, Hancock, and Olofinboba (1997) proposed a combined signal detection and Bayesian analysis of the *posterior* probability of correct alarm response in order to limit false alarms and hence promote effective use.

Equation 31 suggests in addition that automation should be implemented only if the false alarm rate is less than that of the human operator performing the same task.

Other possible simplifications. As with any expected value decision, one needs a priori probabilities, benefits, and costs. With regard to costs, sometimes one can further simplify Equation 28 by assuming that the benefits for doing things right are zero and by being concerned only with costs. In that case one need know only the ratio of $C(\underline{F}|N)$ to $C(\underline{N}|F)$. A ratio is usually much easier to estimate than absolute values.

With regard to the a priori probabilities of failure and normal states of the world, these are already in ratio form, so absolute values are not necessary. The failure modes of human and automation for the task element considered are necessarily a matter for empirical determination.

Finally, no information on benefits or costs is required when the a priori failure probability is either extremely low (~ 0) or extremely high (~ 1), unless benefits or costs are also extremely low or high.

Use of minimax loss criterion. In contrast to selecting human or automation based on whichever has the greatest expected value (benefit minus cost), there is also the minimax loss criterion — that is, selecting human or automation to minimize the worst-case payoff, whatever the value of $p(F)$. From Equation 25 the sure payoff if F occurs is $[p(\underline{F}|F)B(\underline{F}|F) - p(\underline{N}|F)C(\underline{N}|F)]$, whereas if N occurs the sure payoff is $[p(\underline{N}|N)B(\underline{N}|N) - p(\underline{F}|N)C(\underline{F}|N)]$. One of these bracketed terms will be the worse (greater net cost) in the automation case and one in the human case. One selects whichever of human or automation has the better worst case. This is a most conservative criterion, protection against the worst thing that can happen, independent of $p(F)$, which variable is typically not under the system designer's control.

4.7 Competitive Decisions: Games

To this point I have discussed decisions in a benign environment, where the probabilities were determined by nature, so to speak. I now turn briefly to the problem of decision making in a competitive environment — sports, business, war, or personal competition.

Consider the two payoff matrices of Figure A8, and assume that instead of given X and Y probabilities, the X and Y contingencies are determined by an intelligent opponent. The first number in each cell is the corresponding payoff (or penalty) to oneself and the second number is the payoff to the opponent. Each player is trying to maximize his or her payoff. This situation is called *game theory*. I mention only two interesting classes of game situations using these examples. Both examples are for *zero-sum* games, whereby one party's gain is the other party's loss. (This need not be the case; the numbers can take different values.)

Dominating and nondominating strategies. Typically a minimax strategy is employed in zero-sum situations, in which one chooses the option with the least downside risk (the most conservative strategy). For the matrix at the left of Figure A8, one's own minimax is A because it is the best of the worst outcomes, depending on what the opponent does. (For A the worst of -3 and -1 is -3, and for B the worst of -4 and -5 is -5, so the best of -3 and -5 is -3, pointing to A as the minimax). The opponent's minimax is X (the worst of 3 and 4 is 3 for X, the worst of 1 and 5 is 1 for Y, so the best of these is 3, pointing to X).

With these particular, arbitrarily chosen numbers, A happens to be a *dominating* strategy because it is better than B for both X and Y moves of the opponent. The opponent, however, has no dominating strategy, because

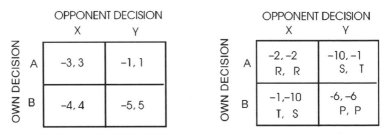

Figure A8. Dominating and nondominating strategy (at left) and prisoner's dilemma (right).

although X is better for the opponent when you choose A, Y is better for the opponent if you choose B.

Prisoner's dilemma. This game is named for the dilemma posed to each of two prisoners in deciding whether to keep silent about the fellow prisoner (cooperate) or provide harmful testimony about that prisoner and gain a lighter sentence (defect). In the matrix at the right of Figure A8, the upper-left cell represents mutual cooperation, in which both keep quiet about the crime (both rewarded, R, with a light sentence). The lower-right cell is mutual defection, in which each testifies against the other (both punished, P). The other two cells illustrate when one testifies (T) against the other and wins a light sentence, whereas the other tries to cooperate but becomes a sucker (S) with a heavy sentence. Thus if you defect (choose B) while your colleague tries to cooperate (chooses X), you lose little and he or she suffers great loss.

The matrix is symmetrical, so it also works the other way around (A, Y). If you both cooperate (A, X), you both suffer only a tiny loss, whereas mutual defection (B, Y) results in moderate punishment. Notice that both players have minimax and dominating strategies (B, Y) — mutual defection. Paradoxically, however, this result (−6 for both players) is significantly worse than the mutual cooperation result (−2 for both).

A prisoner's dilemma is defined by the constraints $T > R > P > S$ and $R > 0.5(S + T)$. This paradoxical situation is not unlike many situations in politics, business, and personal life when each party's effort to obtain an advantage over another party can result in greater cost to both parties than if they cooperate to mutual advantage. Nevertheless, the temptation to defect is so rational! There seems to be a great moral lesson in this simple game, which many researchers have studied and many writers have discussed.

4.8 Signal Detection Theory

I now turn to a theoretical treatment that extends the decision theory described earlier and has wide application in human and machine sensing and attention. Assume an imperfect sensory system, human or machine, given the task of detecting a signal against a background of noise. Assume that sometimes signal plus noise (SN) occurs, otherwise noise alone (N) occurs, and that the probabilities of SN and N are known. As in the previous example, assume there are contingent benefits and costs.

- Let R_{TP} be the benefit for correctly deciding SN when SN is the truth. This situation is called a *true positive, TP.*
- Let C_{FA} be the cost of incorrectly deciding SN when N is the truth. This situation is called a *false alarm, FA.*

- Let C_{MS} be the cost of incorrectly deciding N when SN is true. This situation is called a *miss, MS*.
- Let R_{TN} be the benefit for correctly deciding N when N is true. This situation is called a *true negative, TN*.

Figure A9 presents these values in a payoff matrix or a *reward-cost matrix*.

Finally, let $p(SN)$ be the prior probability of SN and let $p(N)$ be the prior probability of N. Then, $p(N) = 1 - p(SN)$ because the two events are mutually exclusive and collectively exhaustive. To maximize expected reward, one should decide $\underline{SN}$ when the expected value of doing that is greater than the expected value of deciding $\underline{N}$. (As in previous examples, $\underline{SN}$ and $\underline{N}$ are underlined to represent the human decision, and the nonunderlined SN and N represent the actual condition, the truth.)

The expected value criterion is met when the expected value (EV) of deciding $\underline{SN}$ exceeds the expected value of deciding $\underline{N}$:

$$EV(\underline{SN}) > EV(\underline{N}), \tag{32}$$

or when

$$p(SN)R_{TP} + p(N)C_{FA} > p(N)R_{TN} + p(SN)C_{MS}, \tag{33}$$

which assumes the cost values C are negative numbers, or when

$$p(SN)(R_{TP} - C_{MS}) > p(N)(R_{TN} - C_{FA}),$$

or decide $\underline{SN}$ when

$$\frac{p(SN)}{p(N)} > \frac{(R_{TN} - C_{FA})}{(R_{TP} - C_{MS})} \tag{34}$$

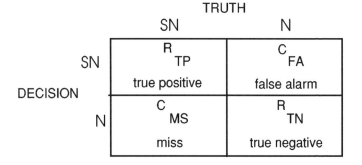

Figure A9. *Payoff matrix for signal detection.*

Now consider that $p(SN)$ and $p(N)$ are prior probabilities before one gets additional evidence e from whatever source. After that bit of new evidence, one has better estimates, $p(SN|e)$ and $p(N|e)$. Thus the after-evidence criterion to optimally decide $\underline{SN}$ is

$$\frac{p(SN|e)}{p(N|e)} > \frac{(R_{TN}-C_{FA})}{(R_{TP}-C_{MS})}. \tag{35}$$

Because $p(SN|e)p(e)=p(e|SN)p(SN)$ and $p(N|e)p(e)=p(e|N)p(N)$, when the two expressions are set in a ratio, the $p(e)$ term drops out, resulting in

$$\frac{p(SN|e)}{p(N|e)} = \frac{p(e|SN)p(SN)}{p(e|N)p(N)} \tag{36}$$

according to Bayes' rule. (See the previous section on Bayes' rule for updating probability estimates based on newly gathered evidence.) When Equation 36 is substituted for the left side of Equation 35 and the terms rearranged with only $p(e|SN)/p(e|N)$ on the left side of the inequality, one obtains the decision criterion in a most useful form: decide $\underline{SN}$ if

$$\frac{p(e|SN)}{p(e|N)} > \frac{p(N)(R_{TN}-C_{FA})}{p(SN)(R_{TP}-C_{MS})}. \tag{37}$$

The left side of Equation 37 is a so-called likelihood ratio for any evidence e, whereas the right side is an a priori constant generally called β in signal detection theory.

The term β represents the threshold at which for larger values one decides $\underline{SN}$ and for smaller values one decides $\underline{N}$. At this threshold one maximizes expected return (based on given contingent costs, benefits, and prior probabilities of SN and N).

Probability densities for evidence. Evidence e can occur in many forms, either discrete or continuous. It is easiest to think of e on an ordinal (not necessarily cardinal) scale, ordered from least to greatest with respect to the ratio $p(e|SN)/p(e|N)$. This, then, is an ordering with respect to the strength of evidence supporting the decision $\underline{SN}$.

With e so ordered, and with probability densities $p(e|N)$ and $p(e|SN)$ plotted on the same e abscissa as in Figure A10, one obtains a very useful characterization of the situation. The density curves — or normalized histograms when one has discrete data — need not be neat bell-shaped curves; they can be any shape. (With Gaussian density functions the afore-mentioned ratio $p(e|SN)/p(e|N)$ is always monotonic with e.)

The density functions are very likely to overlap, so for some values of e there is significant probability density that SN is true and also that N is

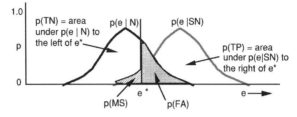

Figure A10. Density functions for $e|N$ and $e|SN$, $e^* = \beta$ for an optimal decision.

true. In Figure A10 one can easily identify the areas indicated for true positive *(TP)*, true negative *(TN)*, false alarm *(FA)*, and miss *(MS)*, the sizes of which represent the corresponding aggregate probabilities for any decision cutoff point e^* (whether or not it equals β, the ideal).

The principal questions are for what values of e to decide $\underline{N}$ and for what values to decide $\underline{SN}$. Equation 37 gives the ideal answer for the given prior probabilities, rewards and costs, independently of the shapes of the probability density curves $p(e|N)$ and $p(e|SN)$. To repeat, at that e where the ratio of the density curves $p(e|SN)/p(e|N) < \beta$, it is best to decide $\underline{N}$, and when $p(e|SN)/p(e|N) > \beta$, it is best to decide $\underline{SN}$.

Relative operating characteristic (ROC). If one considers all values of e^* from very small to very large and cross-plots $p(FA)$ on the abscissa and $p(TP)$ on the ordinate, one gets what is commonly called a *relative operating characteristic* (ROC) curve (see Figure A11). Curve A is an ROC corresponding to the overlap of the two density curves in Figure A10.

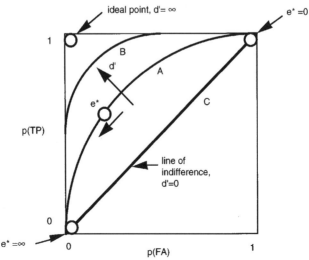

Figure A11. Relative operating characteristic (ROC) curve.

Note that $p(FA)$ is obtained by integrating $p(e|N)$ from e^* to ∞, and $p(TP)$ is likewise obtained by integrating $p(e|SN)$ from e^* to ∞. That means that when e^* is 0, then $p(TP)$ and $p(FA)$ will both be at the maximum value, 1, and when e^* is ∞, then $p(TP)$ and $p(FA)$ will both be 0. Note also that because $p(MS)=1=p(TP)$ and $p(TN)=1-p(FA)$, all four probabilities are represented on the ROC curve as a function of e^*.

Figure A12 illustrates the very happy situation of excellent detectability (discrimination), such that the probability density curves for $p(e|N)$ and $p(e|SN)$ are barely overlapping. The corresponding ROC curve is given by Curve B in Figure A11. This curve comes close to the ideal operating point at the upper-left corner, where $p(TP)=1$ and $p(FA)=0$.

For the unhappy situation of extremely poor detectability (little or no discrimination), where the two density curves have barely any separation and essentially lie on top of each other, there is no e^* that provides a good decision point. Then $p(TP)$ and $p(FA)$ will be essentially equal for all values of e^*. This situation, in the extreme, corresponds to the ROC Curve C in Figure A11.

The number of standard deviations between the means of the two density curves is an accepted measure of discriminability, commonly called d' in signal detection theory. For a lower-left to upper-right diagonal (Curve C, Figure A11) on the ROC $d'=0$, whereas $d'=\infty$ at the upper-left corner.

4.9 Fuzzy Signal Detection

Recently Parasuraman, Masalonis, and Hancock (2000) proposed that both signal (S presented) and response (decide S) can have varying degrees based on various properties of a stimulus or of a response situation and that these can be modeled by fuzzy logic. Accordingly, both the signal and the response are assigned monotonic fuzzy memberships s and r, respectively. Figure A13 gives these authors' example for aircraft separation conflict, which is crisp by FAA rules but to most people is a fuzzy concept — that is, any single separation distance can be partly conflict and partly nonconflict.

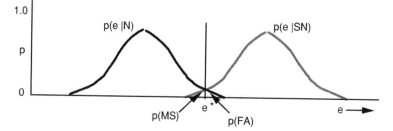

Figure A12. Relative operating characteristic for almost ideal discriminability.

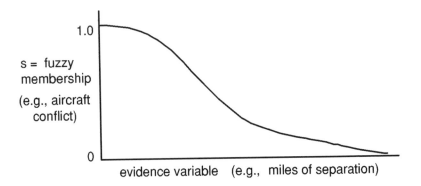

1.0

s = fuzzy
membership
(e.g., aircraft
conflict)

0

evidence variable   (e.g., miles of separation)

Figure A13. Fuzzy membership for aircraft conflict. (From Parasuraman, Masalonis, & Hancock, 2000, adapted with permission from Raja Parasuraman.)

Parasuraman et al. (2000) proposed "implication functions," which exhibit interesting properties. These modifications of conventional or crisp signal detection definitions are given below, along with my own interpretation of the meaning of each. These functions imply that an event can be characterized as simultaneously being in more than one of the four truth table categories.

hit:	$H = \min(s, r)$	the overlap of s by r
miss:	$M = \max(s - r, \ 0)$	the undersupport of s by r
false alarm:	$FA = \max(r - s, \ 0)$	the oversupport of s by r
correct rejection:	$CR = \min(1 - s, \ 1 - r)$	the overlap of absence of s by absence of r

It can be shown that the values of the four terms always sum to one. Parasuraman et al. (2000) went on to define the normalized hit, miss, false alarm, and correct rejection rates as

hit rate:	$HR = \Sigma(H_i)/\Sigma(s_i)$ for $i = 1$ to N trials
miss rate:	$MR = \Sigma(M_i)/\Sigma(s_i)$ for $i = 1$ to N trials
false alarm rate:	$FAR = \Sigma(FA_i)/\Sigma(1 - s_i)$ for $i = 1$ to N trials
correct rejection rate:	$CRR = \Sigma(CR_i)/\Sigma(1 - s_i)$ for $i = 1$ to N trials

As in crisp signal detection theory, given the foregoing definitions, $HR + MR = 1$ and $FAR + CRR = 1$, so the average performance over N trials can be specified by only two terms — for instance, HR and FAR. The parameter d' (in crisp theory the number of standard deviations between Gaussian S and N probability density functions) may be determined as the difference between tabled Z scores on the cumulative Gaussian function corresponding to *HR* and *FAR*. The parameter β in fuzzy theory is the same as in crisp theory and can be determined by the ratio of ordinates of Gaussian probability densities corresponding to *HR* and *FAR*, respectively.

The reader is advised to consult the reference (Parasuraman et al., 2000) for details on how to use the fuzzy signal detection model.

4.10 Calibrating a Decision Maker against Reality

A problem with extracting judgment information from people is that their judgments about the values of variables in the physical world are biased. This problem has a solution, for those judgments can be made more accurate and more useful by a process of *calibration*.

Calibration makes several assumptions:

1. Mental models and judgment processes are relatively stable statistically, so that individual judgment biases about the same or closely related variables in the same or similar situations can be treated in the same way — that is, as a probability density.
2. For any single judgment, the human judge may have trouble considering the shape of the subjective probability density function but could more easily specify his or her "best guess" (median) of the judgment density function for that event. The human judge can also provide estimated fractiles (say, 5th percentile and 95th percentile) on the judgment as a way of specifying how sure he or she is. (Computers, by the way, don't usually specify this when they render estimates, but they should be encouraged to.)
3. Eventually the judgments can be verified against events in the physical world— eventually the truth is known.

Assume that a number of judgments are made that relate to the same or a similar situation. They can be judgments of the same variable at different times or places or judgments of different variables, but all are assumed to involve the same mental model. In this case, except for scaling factors, the biases and degrees of precision of judgments on the different variables are likely to be the same.

In each case the judge is asked to provide not only the best estimate (median) but also the 90% confidence limits. The first three such judgments are represented by the ticks on the top three lines in Figure A14. Eventually the truth about each judgment is determined, and that is recorded for each variable (as the circles in Figure A14). After a number of such comparisons, the results are aggregated on a common normalized scale, where the 90% confidence ticks are the same and the units of each variable lose their meaning. Now the true values form a frequency distri-

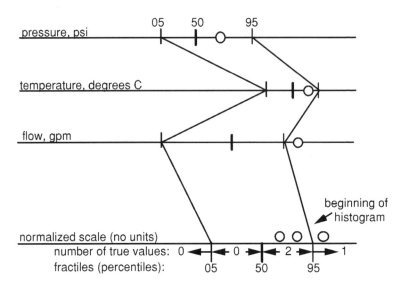

Figure A14. Procedure for calibrating decision maker against reality.

bution, and statistical estimates can be made of the mean and the standard deviation.

If the judge had no biases in either his or her best estimate or 5% and 95% fractile estimates, the median of the true values would lie at the 50% judgment fractile tick, and the 5% and 95% fractiles of the true values histogram (made up from all the circles) would correspond to the 5% and 95% subjective judgment fractiles. The offset of the 50% judgment fractiles relative to the 50% of the true values is the expected bias. If the 90% confidence of the true-value histogram lies outside the 90% subjective judgment ticks, that implies that the judge has more confidence than is warranted. If the 90% confidence of the true-value histogram lies inside the 90% subjective judgment range, that implies that the judge has less confidence than he or she should have in his or her own judgments.

A chi-square statistic adapts easily to testing whether the expected number of true values appears in each of the interfractile intervals. Mendel and Sheridan (1989) extended these ideas for getting calibrated "best measures" from multiple human experts or other information sources.

5. Continuous Control

Control means to make one variable, the *state variable*, which one wants to control, conform to some other variable, a *reference variable*. Often this is to be done in spite of still another variable, a *disturbance variable*, which tends to change the state variable in unpredictable ways. Thus it is more than just decision making; it also includes the sensing and the action taking.

The reference variable can be fixed, and in that case the control is called *regulation*. If the reference variable is changing, the control is called *tracking*. Control has three ingredients: (a) measurement of the state of the variable to be controlled; (b) decision making, based on a *control law*; and (c) actuation to force the controlled process to do what is desired. Control is actually the implementation of "if-then" decision rules, which can be simple rules or more complex sets of rules, and in turn can be either discrete or continuous. In the latter case the decision rules tend to be differential equations.

5.1 On-Off Control

The simplest kind of control is *on-off control*. That is the way the home thermostat usually works. If it is too cold (that is, the measured temperature falls below a set point), the heat source is made to switch on. This step input in heat flow produces a gradually increasing temperature, according to the dynamic equations relating heat flow to temperature. After a few minutes the temperature will reach a second set point, a threshold of being too hot. At that point the heat source switches off, and the temperature cools until the too-cold set point is reached, and the cycle begins all over again.

Some controllers are *on-off-on*, whereby the two *on*s are in opposite directions. If the temperature in the preceding example got too hot, one could switch on an air conditioner, but that would be wasteful. A better example would be the small gas thrusters that steer a spacecraft. When a thruster is on, it produces a constant force in one direction. An opposing thruster produces a constant force in the opposite direction. The spacecraft velocity changes as a function of how long any thruster stays on. If no thruster is active, the spacecraft just coasts at constant velocity. By means of a set of thrusters pointed in different directions, the velocity can be controlled over a wide range in any direction.

5.2 Variables and Transfer Functions

Now I will look more closely at input-output dynamic relations and show how this fits with control that is continuous in time, space, and effort.

Control engineers are fond of using block diagrams, with lines representing variables (like x and y) that change with time, making them $x(t)$ and $y(t)$, and blocks representing input-output relations, typically differential equations. The latter relations are called *transfer operators* or *transfer functions*, and their constants are called *parameters*.

A transfer operator is an operation on a specified variable (an operator in the mathematical sense, not a human operator!) in which a transformation on the input variable results in an output variable. A transfer operator is typically written as a linear polynomial (sums of numeric terms) with coefficients and powers of $D = (d/dt)$, for example,

$$y = [K_0 + K_1 D + K_2 D^2]x, \tag{38}$$

where the terms in brackets constitute the *Heaviside* transfer operator (or linear differential equation operator), x is the input variable, and y is the output variable. Translated, this means $y(t) = K_0$ times $x(t)$, plus K_1 times the time derivative of x, plus K_2 times the second derivative of x. The polynomial transfer function in brackets is often designated by a capital letter, such as G, so that $y = Gx$.

Later I show why the output variable y is usually written as a product of a polynomial function of s and an input variable (i.e., replacing D in Equation 38 with s). This implies that the differential operator terms and the input time function $x(t)$ have been Laplace-transformed by the equation

$$\text{Laplace - transformed} \quad \text{term} = F(s) = \int_0^\infty f(t)e^{-st}dt, \tag{39}$$

where $f(t)$ represents any time function, such as $x(t)$ or $y(t)$, and $F(s)$ represents the corresponding Laplace-transformed variable, usually designated by the corresponding capital letter. The Laplace transform of G, the terms of a linear differential equation, are easy because one need remember only that the Laplace transform of a derivative is s, that of an integration is $1/s$, that of a constant is 1, and so on (tables of these transforms are commonly available in mathematics texts. After Laplace transformation, Equation 38 must be written as

$$Y(s) = [K_0 + K_1 s + K_2 s^2]X(s), \tag{40}$$

and now there is a big advantage because one can simply multiply the transformed operators by the transformed time functions in any analysis. The bracketed term is now called *a transfer function*. Sometimes transfer functions are written in the form

$$y(j\omega) = (K_0 + K_1 j\omega + K_2 \omega^2)x(j\omega), \tag{41}$$

which means that instead of Laplace transforms, Fourier transforms have been used. That is, instead of the Laplace multiplier e^{-st}, the Fourier multiplier $e^{-j\omega t}$ is used in Equation 41. The difference is that the Laplace argument s includes both real and imaginary components, whereas the

Fourier argument $j\omega$ includes only the imaginary, or pure frequency related, component. When analyzing systems in terms of frequency response using, for example, Bode plots (see later section), Equation 41 is implied. Readers interested in the niceties should look at a basic text on control theory or system mathematics. Suffice it for now to say that one should be aware that control engineers imply transformed variables when they write expressions such as Equations 39, 40, or 41, and in using block diagrams they simply multiply the variables on the lines by operators in the boxes.

5.3 Loop Equations for Continuous Negative Feedback Control

Figure A15 represents the control loop of a human *controller* driving a car. The car is the *process* under control. Each of driver and car has its own transfer function, G_c and G_p, respectively. The reference input is $r(t)$, which in the case of car driving is the twists and turns of the center of the lane. The error, $e(t)$, is the difference between $r(t)$ and the actual lateral position of the car, $x(t)$. The manual response of the driver, or the controller's manipulated variable, is $u(t)$. A disturbance, such as a wind gust, is $w(t)$. A simple model of a car is a first-order integrator, $G_p = 1/s$. One can write

$$X(s) = G_p[U(s) + W(s)] = 1/s[U(s) + W(s)] \tag{42}$$

because for any given steering wheel (or car wheel) angle, the car's lateral deviation increases with time, hence the time integral. (Note that this model works only with small angular deviations of the car from the road that are quickly corrected. For large, uncorrected wheel angles, over time the car goes in a circle!)

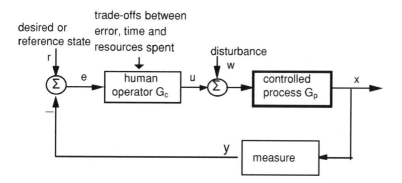

Figure A15. Simple closed loop with human operator as controller.

Negative feedback — namely, the driver's tendency to steer opposite to any lateral deviation — is represented in Figure A15 by the subtraction junction (circle at left with summation and minus sign). Let the driver's transfer function G_c be simply a constant K, which in this case is also the *controller gain*.

Now, dropping the Laplace argument s but retaining the capital letters as a reminder that transformed variables are being used, from the block diagram connections one can write (assuming measurement to be perfect, so that $y = x$)

$$X = G_p(U + W) = G_pG_c(R - X) + G_pW. \tag{43}$$

Collecting terms and rearranging gives the transfer functions for R and W,

$$X = (G_cG_p)R/(1 + G_cG_p) + G_pW/(1 + G_cG_p). \tag{44}$$

Substituting K for G_c and $1/s$ for G_p gives

$$X = (KR/s)/(1 + K/s) + (W/s)/(1 + K/s) = KR/(s + K) \tag{45}$$
$$+ W/(s + K),$$

which has the form of a first-order exponential lag. X/R and X/W are called the *closed-loop transfer functions* because each is a ratio of system output to a system input. Note that as the driver's gain K approaches infinity, the closed-loop transfer function X/R approaches 1, which means the car follows the twists and turns of the road perfectly, while X/W approaches zero, which means the disturbance is completely canceled.

This could also be shown by deriving from Equation 42 the fact that

$$E = R/(1 + G_cG_p) - G_pW/(1 + G_cG_p), \tag{46}$$

which, after substituting $G_c = K$ and $G_p = 1/s$, as before, yields

$$E = sR/(s + K) - W/(s + K). \tag{47}$$

So as gain K gets very large, the error approaches zero.

Thus the simple solution to automation might seem to be negative feedback with as high a gain as possible, to be as sensitive as possible to any error between $r(t)$ and $x(t)$. Unfortunately, it is not so easy, as will be discussed in the next section.

5.4 Stability

A critical requirement of any automatic control is to keep the closed loop from going unstable. What is instability? It is a condition whereby the output gets bigger and bigger, even though the input is constant or even zero. How can that happen?

Consider a chair. If you tilt the chair slightly with two legs off the floor, the center of gravity is still inside all the legs, so if you relax, the chair plops back onto its four legs. However, if you tilt too far, so the center of gravity is outside the four legs, the chair becomes unstable — it does not return to its original equilibrium position but falls over.

In a stable system there is an inherent negative-feedback closed loop: The restoring forces tend to push the chair back in the negative direction from the tilt. When a critical angle of tilt is reached, the force is not restoring and the feedback is positive rather than negative — it tends to make the chair go farther in the same direction and make it fall over.

Consider Figure A15 again. What would make the feedback in car driving turn positive, assuming the driver is not simply perverse? Suppose the driver is sleepy and becomes a bit slow in generating $u(t)$ to follow the rapid twists and turns of a country road, $r(t)$. If the response is slow to the point that it is $180°$ out of phase with the twists and turns (the frequency content of the input signal), the driver is essentially putting in a minus sign in addition to the original minus sign, resulting in a net positive sign: positive feedback! That means that whatever the deviation of the car from the center of the lane, instead of correcting it, the driver only makes it increasingly worse in proportion to the gain. What happens in this case is that the oscillations of the car relative to the road become increasingly worse until the car runs off the road.

(You may not believe this, but I actually tried this experiment with a graduate student after up to 24 hours of sleep deprivation, using a car with dual controls and myself in the other seat. The transfer function got close to $180°$ but not quite. We did not go into the ditch.)

Although this does not normally happen with ordinary cars, it does happen with dual-tandem tractor trailers and large ships, which have a very slow response time and dynamic characteristics (more complex than just $1/s$) that make the response slow relative to the frequency of control signals required.

The formal condition for instability is that the gain must be greater than 1 for any component of energy at the frequency at which phase lag is $180°$.

So control engineering is largely a matter of making the gain large enough that errors are nulled but at the same time shaping the loop dynamics (the transfer function) to avoid instability, which can be exacerbated by large gains as well as by slow dynamics.

A transfer function in general consists of a numerator polynomial in powers of s and a denominator polynomial in powers of s. Depending on their order and complexity, numerator and denominator polynomials can be factored into such terms as $(s+a)$, $(s+b)$, a and b being real numbers, and/or pairs such as $(s+ic)$, $(s+id)$, where i is $\sqrt{-1}$ and ic and id are therefore imaginary numbers. Instability results (numerator/denominator $\to \infty$) when

$$(\text{denominator polynomial}) = 0. \tag{48}$$

Equation 48 is called the *characteristic equation*. Its solutions or roots ($s = -a$, $s = -b$, $s = -ic$, $s = -id$) determine alternate values of s such that denominator $= 0$. These terms are called *poles*. Solutions to numerator $= 0$ are called *zeroes*. Solutions to the corresponding linear differential equations take the form of e^{st} .

All this is useful because the poles and zeroes of the open-loop transfer function can be plotted respectively on a complex plane (horizontal for real component, vertical for imaginary component). There is a technique (you will be spared here) for determining the loci of the roots of the closed-loop transfer function as loop gain K increases from zero (in which case the closed-loop roots are the same as the open-loop poles) to very large values, in some cases converging to the given open-loop zeroes. Such a plot, when it creeps into the right-half plane, tells at what gain the closed-loop system becomes unstable.

5.5 Gain-Phase (Bode) Plots

A frequently used graphical analysis is called a *gain-phase plot* or *Bode plot* (named after Heinrich Bode of the former Bell Laboratories). It is a dual gain-phase plot in that it plots both the open-loop gain $|G_cG_p|$ and the phase of G_cG_p as functions of frequency ω, as in Figure A16. Accurate plots do not have straight lines, but straight-line approximations are convenient when logarithmic scales are used, as in Figure A16. The gain plot is

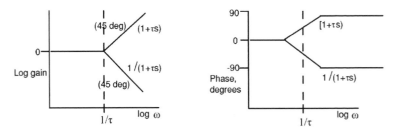

Figure A16. Bode plots for two typical transfer functions. Upper curves are for "lead." Lower curves are for "lag."

the same as what one sees to characterize a sound system. Most processes G_p have gains and phases that decrease with increasing frequency ω, though there may also be bumps that designate resonances, frequencies at which some lasting vibration may occur if the system is excited.

From the Bode plot one can immediately see whether this system, if it were put within a closed loop with negative feedback, is close to instability. One can see if at 180° phase lag the corresponding gain has safely fallen below unity. The degree to which the gain is less than one at 180° phase is called the *gain margin*. One can also check the degree to which the phase is less than 180° when the gain is unity, which is called *phase margin*. These are common indices of "closeness to instability" — that is, a control system factor of safety.

5.6 Optimization

Modern control theory, also known as *optimal control theory*, provides a way to determine a best trade-off among control error, control effort (usually energy), and time. It requires two types of knowledge at the outset:

1. an explicit model (equations or computer simulation) of the controlled process G_p, and
2. an objective function, meaning a statement of relative goodness of system performance as a function of control error e, control effort u, and time T for a transient response to settle.

The model need not be perfect, only good enough to give a fair estimate of $x(t)$ as a function of past $x(t)$ and current control effort. The objective function often takes the form of a weighted sum of quadratic terms, such as $J = K_1 e^2 + K_2 u^2 + K_3 T^2$. Figure A17 shows how the elements of an optimal control system are arranged.

5.7 Fuzzy Control

Linear control such as I have described is what is commonly used to build electromechanical controllers and to characterize human control actions. However, when the processes involved become nonlinear, linear control laws do not work so well. The use of fuzzy logic for control decisions is coming into increasing use in practice. Control theorists, however, do not always like fuzzy control because it does not adapt to neat mathematical stability analysis of the type described earlier.

Fuzzy control might best be illustrated by an example. Consider braking a car and what rules a person might be told to use for various combinations of distance D and relative (closing) velocity V with an object on a collision

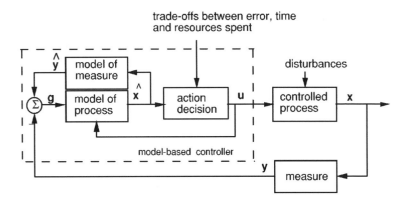

Figure A17. Modern control paradigm, where action u is based on model's estimation of current state x, and process model is continuously updated by nulling the discrepancy g in measurements of modeled and actual states.

course. One can think of a set of fuzzy variables for *D*, such as *very far, far, medium, close,* and *very close,* mapped onto the physical variable, distance in meters. Similarly, imagine a set of variables *slow, medium, fast,* and *very fast* for *V,* mapped onto velocity in meters per second. Any combination of five *D* categories and four *V* categories results in a 5 × 4 matrix. A fuzzy rule specification might simply be a numerical braking force in each cell of the matrix, say, in equivalent kilograms applied to the brake pedal.

In reality, for typical overlapping fuzzy membership functions (see the previous section on fuzzy decision making), any exact numerical *D* and *V* might have memberships on only two of the fuzzy *D* variables and zero on the others, and the same for *V.* Assume, for example, that for a given distance and velocity, the only fuzzy memberships with finite value are *D* = 0.6 (medium) and 0.4 (close) and *V* = 0.3 (fast) and 0.7 (very fast). Figure A18 shows the four salient cells of the 5 × 4 matrix, each of which has a number in boldface that might be the braking force recommended for that combination of fuzzy variables.

Because the AND rule applies in every case (i.e., the car is at *D* distance AND traveling at *V* velocity), one takes the minimum of distance and speed memberships for each cell. Then one takes a weighted average to determine the desired braking force:

$$\text{Force} = \frac{(0.3)\mathbf{3} + (0.3)\mathbf{6} + (0.6)\mathbf{5} + (0.4)\mathbf{10}}{(0.3) + (0.3) + (0.6) + (0.4)} = 7.46 \text{ Kg} \qquad (49)$$

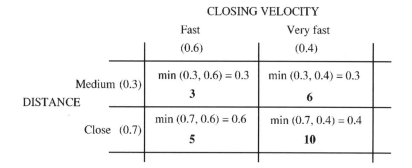

| | CLOSING VELOCITY | |
	Fast (0.6)	Very fast (0.4)
Medium (0.3)	min (0.3, 0.6) = 0.3 **3**	min (0.3, 0.4) = 0.3 **6**
Close (0.7)	min (0.7, 0.6) = 0.6 **5**	min (0.7, 0.4) = 0.4 **10**

Figure A18. Fuzzy membership calculations for salient cells in a fuzzy rule matrix.

Note that both input to and output from the fuzzy controller are cardinal numbers representing multiattribute physical continua. In between are fuzzy rules on a small set of variables. Note also that fuzzy control at this level is simply a way to interpolate within an array of discrete levels for each variable for which the rules are specified.

For a review of fuzzy control, supervision, and fault diagnosis, see Isermann (1998).

5.8 Automation beyond Control Theory

The art of automation goes well beyond control theory. In current practice, for systems of even moderate complexity, the modeling and objective function requirements have often been found to be too restrictive to apply optimal control theory. Most control practice today is still based on classical control theory (described in previous sections) plus a large measure of engineering empiricism. Fuzzy control is becoming more popular because it is robust.

Theory and available computer design packages notwithstanding, control requires having good models of the controlled process (many of which are nonlinear and thus make classical control theory inapplicable). It also requires having good sensors to measure the physical phenomena of interest. Sensors are getting better, smaller, and cheaper but perhaps are the most significant bottleneck to good control. In addition, there are requirements for small, powerful motors and for small, reliable, and fast computers.

Using digital computers in control loops rather than analog circuitry has required profound changes in the thinking of control engineers. As fast as they are, computers have limitations. It takes a few microseconds to sample and hold the voltage of an electric signal. It takes a number of additional microseconds to do the required control processing, effecting the desired G_c transfer function. It takes still more microseconds to convert that digital signal back into an analog signal and amplify it to drive a motor. This whole process, a single control cycle for a single variable, may take one millisecond or more.

If multiple control loops are being handled by a single computer, as is common in robotics, the bandwidth of control may have to be well below 1 kHz. In today's world of control this is not very fast. True, many controllers can happily operate much more slowly (e.g., home heating systems, valves for large volumes of flow, control of vehicles). However, speed is increasing while size and energy requirements and cost are decreasing. A whole branch of control theory has been built up around discretely sampled control and digital approximations of continuous processes.

Appendix B

SAMPLE PROBLEMS AND SOLUTIONS OF NORMATIVE MODELS

1. Problems on Topics Covered in Appendix A

1.1 Supervisory Control

Think of some control system (in a vehicle, industrial or home system, etc.) that is now directly manually controlled but that could be made to be controlled in supervisory fashion. Identify what the planning, teaching, monitoring, intervening, and learning roles of the supervisor might be in terms of the specific variables and context of that system.

1.2 Bayes' Theorem 1

Over a long period, batteries purchased at a local store have been known to be OK 90% of the time. At the moment, all you have available is a cheap battery tester that has indicated failure 5% of the time when used on a battery known for sure to be OK and, when used on a battery known for sure to be bad, has indicated OK 15% of the time. On a series of five tests on a given single battery, the cheap tester indicated OK twice and failure three times. What probability should you best assume that the given battery is OK?

1.3 Bayes' Theorem 2

On three independent tests there is one indication of Condition X and two indications of no X. The prior probability of Condition X is 1%. It is also known that the indications of X are false alarms 10% of the time and misses 1% of the time.

a. Given the test data, what is the probability of X?
b. If the annoyance cost of a false alarm were $1, what would the cost of a missed X have to be for you to act as though X were true?

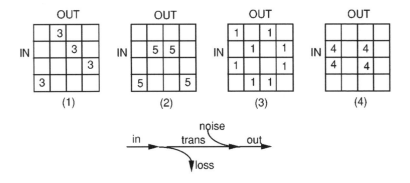

Figure B1. Problem 1.4.

1.4 Information Transmission 1

The cells in the four input-output matrices shown in Figure B1 indicate number of joint events. Sketch four corresponding diagrams such as that shown and indicate on each the input, output, transmitted, equivocation (loss), and noise information in bits.

1.5 Information Transmission 2

Input-output frequency data from an experiment are shown in Figure B2. Considering the capital letters on the matrix as the input set, the lowercase letters as the output set, and the numbers in the cells as the joint occurrences of input-output combinations, determine the input, output, equivocation, noise, and transmitted information ($\log_2 3 = 1.585$, $\log_2 5 = 2.322$).

	a	b	c	d	e
A	1		1		3
B		2	2	1	
C	2		1	1	1
D			1	3	1

Figure B2. Problem 1.5.

1.6 Information Value

Assume that the world has four possible states, A, B, C, and D, occurring with probabilities .2, .3, .1, and .4, respectively. Under various states the return for actions u1, u2, u3, and u4 (including cost of action) is as shown in Figure B3 ($K).

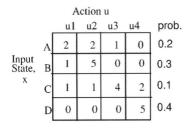

		Action u				prob.
		u1	u2	u3	u4	
	A	2	2	1	0	0.2
Input State, x	B	1	5	0	0	0.3
	C	1	1	4	2	0.1
	D	0	0	0	5	0.4

Figure B3. Problem 1.6.

a. What action would you take if you had no better prediction of state than the prior odds? How much on average would you earn per action?

b. What is the expected return from hiring a clairvoyant able to predict the future perfectly and deciding each action based on the clairvoyant's advice, assuming he or she charges $2K for this service? What are the odds of finding such a person?

1.7 Utility

Use the Von Neumann utility elicitation procedure (estimating certainty equivalent indifference to a 50-50 lottery of maximum and minimum values for a given range) to do the following:

a. Measure the utility of a friend for a gift of money in the range $0 to $1,000,000. Get three intermediate points.

b. Measure the same person's utility for the weight of a laptop computer (gift). Get three intermediate points between 0 and 12 pounds.

Show the results on a graph of utility versus physical attribute in each case.

1.8 Fuzzy Logic 1

You are given the fuzzy membership functions shown in Figure B4 and two rules: If pressure is LOW or temperature is COLD, turn on the heat. If pressure is HIGH and temperature is HOT, do nothing.

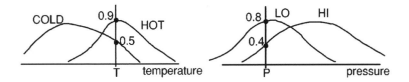

Figure B4. Problem 1.8.

a. What would you decide for the given combination P and T? Show the steps of your logic.
b. Would your decision change if the conjunction in the second rule were "or"?

1.9 Fuzzy Logic 2

The foreman of a small crew fighting a forest fire has been given instructions (the two heuristic rules shown in Figure B5) for when to order his crew to retreat. His fuzzy memberships for the terms used by his boss when transformed against physical size and rate of spread are as shown in the figure. Numerical measurements given by a helicopter observer put the size and rate as indicated on the physical scale (abscissa) for each variable. What decision should the foreman make?

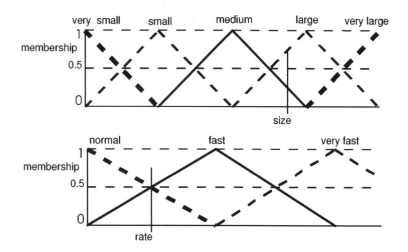

Figure B5. Problem 1.9.

1.10 Decision under Uncertainty with Different Criteria 1

For the payoff matrix shown in Figure B6 (for condition x and action u):

a. What is the minimax (minimax loss) decision, and what is its expected payoff?
b. Based on an expected value criterion, what is the value of having prior information about which *x* will occur (over just knowing the *x* probabilities)?

	x1	x2
probability =	0.3	0.7
u1	8	3
u2	0	10

Figure B6. Problem 1.10.

1.11 Decision under Uncertainty with Different Criteria 2

Given mutually exclusive states *x*, *y*, and *z*, along with probabilities of occurrence and decision consequences as shown in Figure B7, what decision (A, B, or C) would yield the greatest expected value? What would be the most conservative (minimax) decision? What would the obsessive gambler decide?

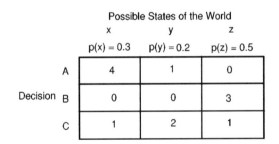

		Possible States of the World		
		x	y	z
		p(x) = 0.3	p(y) = 0.2	p(z) = 0.5
	A	4	1	0
Decision	B	0	0	3
	C	1	2	1

Figure B7. Problem 1.11.

1.12 Signal Detection 1

For the probability density functions shown in Figure B8:

a. Draw the ROC curve with $p(\text{decide } S+N|S+N)$ on the vertical and $p(\text{decide } S+N|N)$ on the horizontal axes.
b. Considering Points A, B, C, D, and E as alternative criterion points for deciding between N and $(S + N)$, identify these points on the ROC.
c. What is the slope of the ROC at Point C?
d. What are the numerical values on the ROC for Point C, and what is the slope at that point?
e. If $p(S+N)=.75$, what is the value of $[(R_{TN}-C_{FA})/(R_{TP}-C_{MS})]$ for C as a criterion?

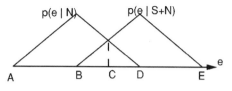

Figure B8. Problem 1.12.

1.13 Signal Detection 2

An experimental participant in a discrimination task was asked to say "one" when he was convinced A was true, "two" when he thought A was true but was not sure, "three" when he thought B was true but was not sure, and "four" when he was convinced B was true. When A was actually true, he said "one" once, "two" twice, "three" once, and "four" not at all. When B was actually true he said "one" not at all, "two" once, "three" twice, and "four" once. Draw the ROC curve representing his discriminability. What evidence is there that he is doing better than guessing?

1.14 Calibration of a Decision Aid

A weather forecaster makes the predictions shown in Figure B9 of wind (km per hour), ice buildup (cm), snow (cm), and low temperature (degrees centigrade) for a particular winter day, giving his 10% and 90% confidence limits and his median predictions. The actual measures turned out as shown in the box to the right in the figure. What evidence is there that the forecaster is biased? What evidence is there that he has too little or too much confidence in his estimates?

	10%	50%	90%	After-the-fact
wind	0	15	40	18
ice buildup	0	1	4	2
snow	2	4	12	5
extreme temperature	0	-8	-20	-12

Figure B9. Problem 1.14.

1.15 Control 1

If the controlled process is an integrator ($1/Ts$) and the controller is a constant K:

a. What is the transfer function from reference input to process output?
b. What is the transfer function from reference input to error?

1.16 Control 2

a. Is the system depicted in the Bode plot in Figure B10 (assuming the lower gain plot) stable? Why or why not?
b. What is the gain margin? What is the phase margin?
c. If the gain were increased by a factor of 10, what can you say about stability? Why?

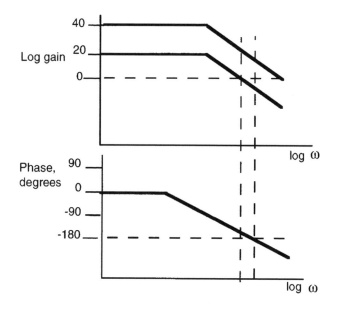

Figure B10. Problem 1.16.

2. Solutions to Problems Stated in Appendix B

2.1 Supervisory Control
(No unique answer)

2.2 Bayes' Theorem 1
Prior probabilities: $p_0(OK)=0.9$, so $p_0(\text{Not OK})=0.1$
Likelihood information:

It is given that $p(\underline{\text{Not OK}}|OK)=0.05$, so $p(\underline{OK}|OK)=0.95$ (underline indicates reading of the tester) and it is given that $p(\underline{OK}|\text{Not OK})=0.15$, so $p(\underline{\text{Not OK}}|\text{Not OK})=0.85$.

Bayes theorem:

$$\frac{p(\text{OK}|\text{all data})}{p(\text{Not OK}|\text{all data})}$$

$$=\left[\frac{p(\text{OK}|\text{OK})}{p(\text{OK}|\text{Not OK})}\right]^2\left[\frac{p(\text{Not OK}|\text{OK})}{p(\text{Not OK}|\text{Not OK})}\right]^3\frac{p_0(\text{OK})}{p_0(\text{Not OK})}$$

$$=\left[\frac{0.95}{0.15}\right]^2\left[\frac{0.05}{0.85}\right]^3\frac{0.9}{0.1}=0.0734.$$

Since p (Not OK)$=1-p(\text{OK})$, $p(\text{OK})=\dfrac{0.0734}{1.0734}=0.0684$

2.3 Bayes Theorem 2

$p_0(x)=0.01$; $p_0(\text{not } x)=0.99$

$p(\text{FA})=p(\underline{x}|\text{not } x)=0.10$; $p(\text{not } x|\text{not } x)=0.90$

$p(\text{MS})=p(\text{not } x|x)=0.01$; $p(\underline{x}|x)=0.99$

$$\frac{p(x|\text{data})}{p(\text{not } x|\text{data})}=\frac{p(\underline{x}|x)}{p(\underline{x}|\text{not } x)}\left[\frac{p(\text{not } x|x)}{p(\text{not } x|\text{ not } x)}\right]^2\frac{p_0(x)}{p_0(\text{not } x)}$$

$$=\frac{0.99}{0.10}\left[\frac{0.01}{0.90}\right]^2\frac{0.01}{0.99}=\frac{1}{81000}$$

a. $p(x|\text{data})-1/81001=0.0000123$
b. Cost of a false alarm= \$1, so the cost of a miss would have to be \$81000 for one to act as though x were true, since there is a trade-off between $[p(x)C_{\text{MS}}]$ and $[p(\text{not } x)C_{\text{FA}}]$.

2.4 Information Transmission 1
(See Figure B11.)

2.5 Information Transmission 2
Note: Because all terms are combinations of digits 1, 2, 3, 4, and 5, fractions can be factored, e.g., $1/(0.15) = (20)/(3) = (4)(5)/(3)$, which can be treated as log 4 + log 5 − log 3. (See Figure B12 on page 240.)

INPUT $H(\text{in})=\sum_i p_i\log_2(1/p_i)=4(0.25)(1/0.25)=2.00$ bits

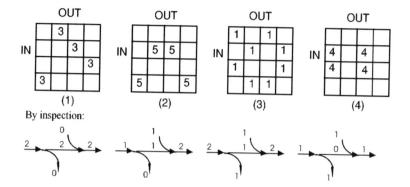

Figure B11. *Answer to Problem 1.4.*

OUTPUT $H(\text{out}) = \sum_j p_j \log_2(1/p_j) = [0.15\log_2(1/0.15)]$

$$+ [0.1 \log_2(1/0.1)] + [3(0.25)\log_2(1/0.25)]$$

$$= 2.24 \text{ bits}$$

TRANSMITTED $H(x:y) = \sum_i \sum_j p_{jj}\log_2[(p_{j|j})/p_i]$

(considering separately: 1 in column a, 2 in a, 2 in b, 1's in cde, 2's in cde, and 3's in cde)

$1/20 \log_2[1/3 / 1/4] + 2/20 \log_2[2/3 / 1/4] + 2/20 \log_2[2/2 / 1/4]$

$$+ 7/20 \log_2[1/5 / 1/4] + 2/20 \log_2[2/5 / 1/4] + 6/20 \log_2[3/5 / 1/4]$$

$$= (0.02075) + (0.1415) + (0.2000) - (0.1127) + (0.0678) + (0.3789)$$

$$= 0.70 \text{ bits}$$

EQUIVOCATION $= H(\text{in}) - H(x:y) = 1.30$ bits

NOISE $= H(\text{out}) - H(x:y) = 1.54$ bits

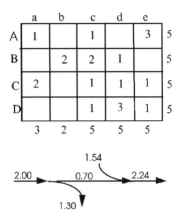

Figure B12. Answer to Problem 1.5.

2.6 Information Value

	Expected values for:				Expected values for clairvoyant
	u1	u2	u3	u4	
A	0.4	0.4	0.2	0	0.4
B	0.3	1.5	0	0	1.5
C	0.1	0.1	0.4	0.2	0.4
D	0	0	0	2.0	2.0
	0.8	2.0	0.6	2.2	4.3

a. Do u4, get 2.2 return average.
b. Get $4.3 - 2.0 = 2.3$. The chance of finding such a person is poor!

2.7 Utility
(No unique answer)

2.8 Fuzzy Logic 1

a. (LO or COLD) = max (0.8, 0.5) = 0.8
   (HI and HOT) = min (0.4, 0.9) = 0.4
   so turn on heat.
b. (HI or HOT) = max (0.4, 0.9) = 0.9
   so leave alone.

2.9 Fuzzy Logic 2
(Medium or Large) = max (0.25, 0.75) = 0.75
(Fast or Very Fast) = max (0.5, 0.5) = 0.5
and of both = min (0.75, 0.5) = 0.5
(Medium or Less) = max (0.25, 0) = 0.25
Normal = 0.5
and of both = min (0.75, 0.5) = 0.25
0.5 > 0.25; therefore first rule holds: retreat.

2.10 Decision under Uncertainty with Different Criteria 1

a. Minimax is u1 (the best of the worst possibilities). Expected value of u1 = (0.3) 8 + (0.7) 3 = 4.5
b. Expected value of knowing ahead = (0.3) 8 + (0.7) 10 = 9.4.
   Expected value of knowing on a probabilistic basis is
   (0.3) 8 + (0.7) 3 = 4.5 for u1, and
   0 + (0.7) 10 = 7.0 for u2,
   so choose u2
   Information value = 9.4 – 7.0 = 2.4

2.11 Decision under Uncertainty with Different Criteria 2

	Expected	Minimax	Maximum
A	1.4	0	4
B	1.5	0	3
C	1.4	1	2
Best action	B	C	A

2.12 Signal Detection 1
(See Figure B13.)
(a and b): The area under $p(e|N)$ to the right of C = 1/8. Therefore:

(c and d): At point C the slope $= \dfrac{p(e|S+N)}{p(e|N)} = 1$

$$\frac{p(e|S+N)}{p(e|N)} = \frac{p(N)}{p(S+N)}\left[\frac{R_{TN}-C_{FA}}{R_{TP}-C_{MS}}\right]$$

$$1 = \frac{0.25}{0.75}\left[\frac{R_{TN}-C_{FA}}{R_{TP}-C_{MS}}\right]$$

$$\left[\frac{R_{TN}-C_{FA}}{R_{TP}-C_{MS}}\right] = 3$$

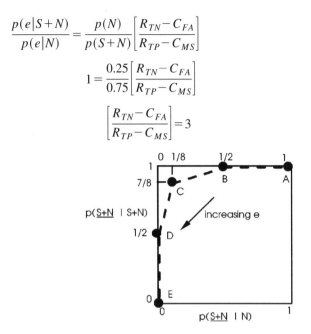

Figure B13. Answer to Problem 1.12.

2.13 Signal Detection 2
(See Figure B14.)

For a criterion (cut-off) at (or just to the right of) each d on the density functions there are two ways to draw the ROC (where d values are indicated on each):

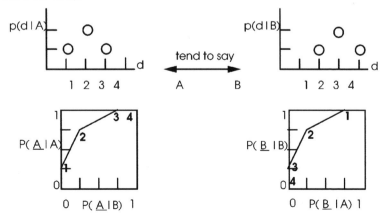

Figure B14. Answer to Problem 1.13.

d	p(A|A)	p(A|B)	p(B|B)	p(B|A)	
1	0.25	0	1	0.75	Note:
2	0.75	0.25	0.75	0.25	P(A|A) + P(B|A) = 1
3	1	0.75	0.25	0	P(A|B) + P(B|B) = 1
4	1	1	0	0	

Evidence is that ROC is closer to ideal point than diagonal (line of guessing)

2.14 Calibration of a Decision Aid
(The scales in Figure B15 are oriented so that bad news is always to the right.)

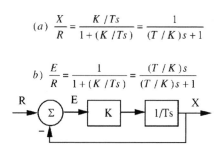

One can conclude:

1) There is a slight bias on the high side.

2) Variability is very small, so the decision-maker has too little confidence in his own judgment of the true values.

Figure B15. Answer to Problem 1.14.

2.15 Control 1
(See Figure B16.)

$$(a) \quad \frac{X}{R} = \frac{K/Ts}{1+(K/Ts)} = \frac{1}{(T/K)s+1}$$

$$b) \quad \frac{E}{R} = \frac{1}{1+(K/Ts)} = \frac{(T/K)s}{(T/K)s+1}$$

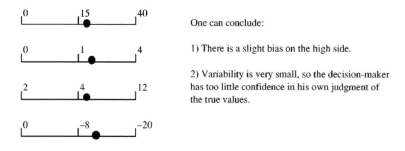

Figure B16. Answer to Problem 1.15.

2.16 Control 2
(See Figure B17.)

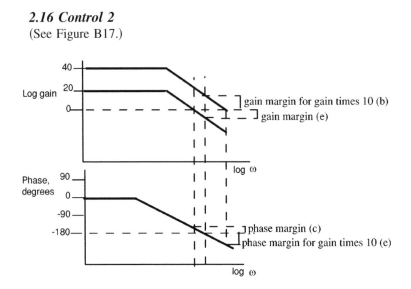

Figure B17. Answer to Problem 1.16.

a. System is stable, since when log gain equals zero (gain equals one), the phase has not yet reached 180°, or, alternatively, when phase has reached 180°, the gain is less than one.

b. Gain margin is the factor by which the gain is less than one at 180° phase lag. Phase margin is the factor by which phase shift is less than 180° when gain is one.

e. When gain is multiplied by 10 (log gain has increased by 20), gain margin and phase margin are both on unstable side (see diagram above).

REFERENCES

Adams, M. J., Tenney, Y. J., & Pew, R. W. (1995). Situation awareness and the cognitive management of complex systems. *Human Factors, 37*, 85–104.

Adelman, L., Cohen, M., Bresnick, T. A., Chinnis, J. O., & Laskey, K. B. (1993). Real-time expert system interfaces, cognitive processes, and task performance: An empirical assessment. *Human Factors, 35*, 243–261.

Arkes, H. R., & Blumer, C. (1985). The psychology of sunk cost. *Organizational Behavior and Human Decision Processes, 35*, 124–140.

Arrow, K. J. (1963). *Social choice and individual values* (2nd ed.). New York: Wiley.

Askey, S. Y. (1995). *Design and evaluation of decision aids for control of high-speed trains: Experiments and a model*. Ph.D. thesis, Massachusetts Institute of Technology, Cambridge, MA.

Askren, W. B., & Regulinski, T. L. (1969). Quantifying human performance for reliability analysis of systems. *Human Factors, 11*, 393–396.

Bainbridge, L. (1983). Ironies of automation. *Automatica, 19*, 775–779.

Bainbridge, L. (1987). Ironies of automation. In J. Rasmussen, K. D. Duncan, & L. Leplat (Eds.), *New technology and human error* (pp. 271–284). Chichester, England: Wiley.

Bainbridge, L. (1997). The change in concepts needed to account for human behavior in complex dynamic tasks. *IEEE Transactions on Systems, Man and Cybernetics, Part A: Systems and Humans*, 351–359.

Barfield, W., & Dingus, T. A. (1998). *Human factors in intelligent transportation systems*. Mahwah, NJ: Erlbaum.

Barfield, W., & Furness, T. A. (1995). *Virtual environments and advanced interface design*. New York: Oxford University Press.

Bar Hillel, M. (1973). On the subjective probability of compound events. *Organizational Behavior and Human Performance, 9*, 396–406.

Billings, C. E. (1997). *Aviation automation: The search for a human-centered approach*. Mahwah, NJ: Erlbaum.

Birmingham, H. P., & Taylor, F. V. (1954). A design philosophy for man-machine control systems. *Proceedings of the Institute of Radio Engineers, 42*, 1748–1758.

Bisantz, A. M., Kirlik, A., Gay, P., Phipps, D. A., Walker, N., & Fisk, A. D. (2000). Modeling and analysis of a dynamic judgment task using a lens model approach. *IEEE Transactions on Systems, Man and Cybernetics, Part A: Systems and Humans, 30*, 605–616.

Bisseret, A. (1970). Memoire operationalle et structure du travail. *Bulletin de Psychologie, 24*, 280–294.

Boff, K., & Lincoln, J. (1988). *Engineering data compendium*. Wright-Patterson Air Force Base, OH: Armstrong Aerospace Medical Research Laboratory.

Booher, H. R. (1990). *MANPRINT: An approach to systems integration*. New York: Van Nostrand Reinhold.

Boy, G. A. (1998). *Cognitive function analysis*. Stamford, CT: Ablex.

Brooks, T. L. (1979). *SUPERMAN: A system for supervisory manipulation and the study of human-computer interactions*. Unpublished S.M. thesis, Massachusetts Institute of Technology, Cambridge, MA.

Brunswik, E (1956). Representative design and probabilistic theory in a functional psychology. *Psychological Review, 62,* 193–217.

Buckner, D. N., & McGrath, J. J. (1963). *Vigilance: A symposium.* New York: McGraw-Hill.

Burdea, G. C. (1996). *Force and touch feedback for virtual reality.* New York: Wiley.

Cacciabue, P. C. (1997). A methodology of human factors analysis for system engineering: Theory and applications. *IEEE Transactions on Systems, Man and Cybernetics, Part A: Systems and Humans, 27,* 325–339.

Carbonell, J. (1966). A queuing model of many-instrument visual sampling. *IEEE Transactions on Human Factors in Electronics, HFE-7,* 57–164.

Chapanis, A., Garner, W., & Morgan, C. (1949). *Applied experimental psychology.* New York: Wiley.

Charny, L., & Sheridan, T. B. (1989). Adaptive goal setting in tasks with multiple criteria. In *Proceedings of the 1989 International Conference on Cybernetics and Society.*

Chen, S., Sheridan, T. B., Kusunoki, H., & Komoda, N. (1995). Car following measurements, simulations, and a proposed procedure for evaluating safety. In *Proceedings of the IFAC Symposium on Analysis, Design and Evaluation of Man-Machine Systems* (pp. 603–608). London: Pergamon.

Christoffersen, K., Hunter, C. N., & Vicente, K. J. (1997). A longitudinal study of the effects of interface design on fault management performance. *International Journal of Cognitive Ergonomics, 1*(1), 1–24.

Cook, R. I., & Woods, D. D. (1996). Adapting to new technology in the operating room. *Human Factors, 38,* 593–613.

Cook, T. D., & Campbell, D. T. (1978). *Quasi experimentation: Design and analysis issues for field settings.* Boston: Houghton Mifflin.

Cooksey, R. W. (1996). *Judgment analysis: Theory, methods and applications.* San Diego: Academic.

Corker, K., Pisanich, G., & Bunzo, M. (1997). Empirical and analytical studies of human/automation dynamics in airspace management for free flight. In *Proceedings of the 10th International Confederation of European Aerospace Societies Conference on Free Flight* (pp. 115–120). Amsterdam: Confederation of European Aerospace Societies.

Craik, K. J. W. (1947a). Theory of the human operator in control systems: 1. The operator as an engineering system. *British Journal of Psychology, 38,* 56–61.

Craik, K. J. W. (1947b). Theory of the human operator in control systems: 2. Man as an element in a control system. *British Journal of Psychology, 38,* 142–148.

Curry, R. E., & Gai, E. G. (1976). Detection of random process failures by human monitors. In T. B. Sheridan & G. Johannsen (Eds.), *Monitoring behavior and supervisory control* (pp. 205–220). New York: Plenum.

Dalkey, N., & Helmer, O. (1963). An experimental application of the Delphi method to the use of experts. *Management Science, 9*(3), 458–467.

Das, H. (1989). *Kinematic control and visual display of redundant teleoperators.* Ph.D. thesis, Massachusetts Institute of Technology, Cambridge, MA.

Degani, A., Shafto, M., & Kirlik, A. (1999). Models in human-machine systems: Constructs, representation and classification. *International Journal of Aviation Psychology, 9,* 125–138.

Degani, A., & Wiener, E. L. (1997). Procedures in complex systems: The airline cockpit. *IEEE Transactions on Systems, Man and Cybernetics, Part A: Systems and Humans, 27,* 302–311.

Dekker, S., & Hollnagel, E. (1999). *Coping with computers in the cockpit.* Aldershot, England: Ashgate.

DeSola Pool, I. (1973). *Talking back: Citizen feedback and cable technology.* Cambridge: MIT Press.

Dowell, J., & Long, J. (1998). Conception of the cognitive engineering design problem. *Ergonomics, 41,* 126–139.

Draper, J. V., Kaber, D. B., & Usher, J. M. (1998). Telepresence. *Human Factors, 40,* 354–375.

Durlach, N. I., & Mavor, A. S. (Eds.). (1995). *Virtual reality.* Washington, DC: National Academy Press.

Eby, D. W., & Kostyniuk, L. P. (1999). An on-the-road comparison of in-vehicle navigation assistance systems. *Human Factors, 41,* 295–311.

Edwards, W. (1954). The theory of decision making. *Psychological Bulletin, 41,* 380–417.

Edwards, W. (1968). Conservatism in human information processing. In B. Kleinmutz (Ed.), *Formal representation of human judgment* (pp. 17–52). New York: Wiley.

Einhorn, H. J., & Hogarth, R. M. (1978). Behavioral decision theory: Processes of judgment and choice. *Annual Review of Psychology, 32,* 53–88.

Einstein, A., & Infeld, L. (1942). *The evolution of physics.* New York: Simon and Schuster.

Ellis, S. R. (1991). *Pictorial communication in virtual and real environments.* London: Taylor & Francis.

Embry, D. E. (1976). *Human reliability in complex systems: An overview* (Report NCSR-R10). Warrington, England: National Center of Systems Reliability, UK Atomic Energy Authority.

Endsley, M. R. (1995). Measurement of situation awareness in dynamic systems. *Human Factors, 37,* 65–84.

Endsley, M., & Kaber, D. B. (1999). Level of automation effects on performance, situation awareness and workload in a dynamic control task. *Ergonomics, 42,* 462–492.

Endsley, M., & Strauch, B. (1997). Automation and situation awareness: The accident at Cali, Columbia. In *Proceedings of the International Symposium on Aviation Psychology* (pp. 877–881). Columbus: Ohio State University.

Engelberger, J. (1981). *IFAC* (International Federation of Automatic Control) *Newsletter,* No. 2, March, p. 4.

Evans, J. B. T. (1989). *Bias in human reasoning: Causes and consequences.* Mahwah, NJ: Erlbaum.

Ferrell, W. R. (1965). Remote manipulation with transmission delay. *IEEE Transactions on Human Factors in Electronics HFE-6* (1).

Ferrell, W. R., & Sheridan, T. B. (1967). Supervisory control of remote manipulation. *IEEE Spectrum, 4*(10), 81–88.

Fischhoff, B. (1975). Hindsight = foresight: The effect of outcome knowledge on judgment under uncertainty. *Journal of Experimental Psychology: Human Perception and Performance, 1,* 288–299.

Fischhoff, B., Slovic, P., & Lichtenstein, S. (1977). Knowing with certainty: The appropriateness of extreme confidence. *Journal of Experimental Psychology: Human Perception and Performance, 3,* 552–564.

Fitts, P. M. (Ed.). (1951). *Human engineering for an effective air navigation and traffic control system.* Washington, DC: National Research Council.

Fitts, P. M. (1954). The information capacity of the human motor system in controlling the amplitude of movement. *Journal of Experimental Psychology, 47,* 381–391.

Flach, J. M. (1995). Situation awareness: Proceed with caution. *Human Factors, 37,* 149–157.

Flach, J. M. (1998). Cognitive systems engineering: Putting things in context. *Ergonomics, 41,* 163–167.

Flach, J. M., Vicente, K., Tanabe, F., Monta, K., & Rasmussen, J. (1998). An ecological approach to interface design. In *Proceedings of the Human Factors and Ergonomics Society 42nd Annual Meeting* (pp. 295–299). Santa Monica, CA: Human Factors and Ergonomics Society.

Fromm, E. (1995). *Escape from freedom.* New York: Holt.

Fuld, R. B. (2000). The fiction of function allocation. *International Journal of Human-Computer Studies, 52,* 217–233.

Gibson, J. J. (1979). *The ecological approach to visual perception.* Boston: Houghton Mifflin.

Gilmore, W. E., Gertman, D. I., & Blackman, H. S. (1989). *User-computer interfaces in process control.* Boston: Academic.

Green, D. M., & Swets, J. A. (1966). *Signal detection theory and psychophysics.* New York: Wiley.

Greenberg, S. (1991). An annotated bibliography of computer supported cooperative work. In S. Greenberg (Ed.), *Computer-supported cooperative work and groupware.* London: Academic.

Grice, H. P. (1975). Logic and conversation. In P. Cole & J. Morgan (Eds.), *Syntax and semantics: Speech acts* (Vol. 3). New York: Academic.

Hall, R. E., Samanta, P. K., & Swoboda, A. L. (1981, January). *Sensitivity of risk parameters to human errors in reactor safety study for a PWR* (Brookhaven National Lab Report 51322 NUREG/CR-1879). Long Island, NY: Brookhaven National Lab.

Hancock, P. A., & Desmond, P. A. (Eds.). (2001). *Stress, workload and fatigue.* Mahwah, NJ: Erlbaum.

Hancock, P. A., & Scallen, S. F. (1996). The future of function allocation. *Ergonomics in Design, 4*(4), 24–29.

Hardin, G. (1968, December). The tragedy of the commons. *Science, 162,* 1243–1248.

Hart, S. G., & Sheridan, T. B. (1984). Pilot workload, performance and aircraft control automation. In *Proceedings of NATO/AGARD Conference 371, Human Factors Considerations in High Performance Aircraft* (18.1–18.12). Neuilly-sur-Seine, France: NATO.

Heidegger, M. (1977). *Being and time* (J. Macquarrie & E. Robinson, trans.). San Francisco: Harper Collins. (Originally published 1927.)

Helander, M. G., Landauer, T. K., & Prabhu, P. V. (Eds.). (1997). *Handbook of human-computer interaction.* Amsterdam: Elsevier.

Helmreich, R. L., & Merritt, A. C. (1998). *Culture at work in aviation and medicine.* Brookfield, VT: Ashgate.

Hick, W. E. (1952). On the rate of gain of information. *Quarterly Journal of Experimental Psychology, 4,* 11–26.

Hoc, J.-M. (2000). From human-machine interaction to human-machine cooperation. *Ergonomics, 43,* 833–843.

Hoc, J.-M., Cacciabue, P. C., & Hollnagel, E. (Eds.). (1995). *Expertise and technology: Cognition and human-computer cooperation.* Mahwah, NJ: Erlbaum.

Hogarth, R. M., & Einhorn, H. J. (1992). Order effects in belief updating: The belief adjustment model. *Cognitive Psychology, 25,* 1–55.

Hollnagel, E., & Woods, D. D. (1983). Cognitive systems engineering: New wine in new bottles. *International Journal of Man-Machine Studies, 18,* 583–600.

Howard, R. A. (1966). Information value theory. *IEEE Transactions on Systems Science and Cybernetics, SSC-2,* 22–26.

Hursch, C .J., Hammond, K. R., & Hursch, J. L. (1964). Some methodological considerations in multiple-cue probability studies. *Psychological Bulletin, 71,* 42–60.

Inagaki, T. (1999). Situation-adaptive autonomy for time-critical takeoff decisions. *International Journal of Modeling and Simulation, 19*(4), 179–183.

Isermann, R. (1998). On fuzzy logic applications for automatic control, supervision, and fault diagnosis. *IEEE Transactions on Systems, Man and Cybernetics, Part A: Systems and Humans, 28*, 221–235.

Johnson-Laird, P. N. (1993). *Human and machine thinking.* Mahwah, NJ: Erlbaum.

Jones, P. M., & Jasek, C. A. (1997). Intelligent support for activity management (ISAM): An architecture to support distributed supervisory control. *IEEE Transactions on Systems, Man and Cybernetics, Part A: Systems and Humans, 27*, 274–288.

Jordan, N. (1963). Allocation of functions between man and machines in automated systems. *Journal of Applied Psychology, 47*, 161–165.

Kaber, D. B., & Riley, J. M. (1999). Adaptive automation of a dynamic control task based on workload assessment through a secondary monitoring task. In M. W. Scerbo & M. Mouloua (Eds.), *Automation technology and human performance: Current research and trends* (pp. 129–133). Mahwah, NJ: Erlbaum.

Kahneman, D., Slovic, P., & Tversky, A. (1982). *Judgment under uncertainty: Heuristics and biases.* New York: Cambridge University Press.

Kahneman, D., & Tversky, A. (1982). The psychology of preferences. *Scientific American, 246*(1), 160–173.

Kalman, R. E. (1960). A new approach to linear filtering and prediction problems. *Journal of Basic Engineering, Transactions of the American Society of Mechanical Engineers, 82D*, 33–45.

Kantowitz, B., & Sorkin, R. (1987). Allocation of functions. In G. Salvendy (Ed.), *Handbook of human factors* (pp. 365–369). New York: Wiley.

Kappel, G. (1992). *Design and analysis: A researcher's handbook* (3rd ed.). Englewood Cliffs, NJ: Prentice-Hall.

Keeney, R. L., & Raiffa, H. (1976). *Decisions with multiple objectives.* New York: Wiley.

Kirlik, A. (1993). Modeling strategic behavior in human-automation interaction: Why an "aid" can (and should) go unused. *Human Factors, 35*, 221–242.

Kirwan, B., & Ainsworth, L. K. (1992). *A guide to task analysis.* London: Taylor & Francis.

Klein, G. A. (1989). Recognition-primed decisions. In W. B. Rouse (Ed.), *Advances in man-machine systems research* (Vol. 5, pp. 47–92). Greenwich, CT: JAI Press.

Kleinman, D. L., Baron, S., & Levison, W. H. (1970). An optimal control model of human response: Part I. *Automatica, 63*, 357–369.

Kohn, L. T., Corrigan, J. M., & Donaldson, M. S. (Eds.) (1999). *To err is human. Building a safer health system.* Washington, DC: National Academy Press.

Kozinsky, E., Pack, R., Sheridan, T., Vreuls, D., & Seminara, J. (1982, November). *Performance measurement system for training simulators* (Report NP-2719). Palo Alto, CA: Electric Power Research Institute.

Kreith, F. (1997). *The CRC handbook of mechanical engineering.* Boca Raton, FL: CRC Press.

Kuchar, J. K. (1996). Methodology for alerting system performance evaluation. *Journal of Guidance, Control and Dynamics, 19*, 438–444.

Kuchar, J. K., & Hansman, R. J. (1993, September). *Part-task simulator evaluations of advanced terrain displays* (Paper SAE-9932570, SAE Aerotech '93). Warren, MI: Society of Automotive Engineers.

Lanzilotta, E. J., & Sheridan, T. B. (1997). *Safety of high-speed ground transportation systems, human factors phase IV: Dynamic risk estimation using the safety state model.* Cambridge, MA: Volpe National Transportation Systems Center.

Lee, J., & Moray, N. (1992). Trust, control strategies and allocation of function in human-machine systems. *Ergonomics, 35*, 1243–1270.

Lee, J., & Moray, N. (1994). Trust, self-confidence, and operators' adaptation to automation. *International Journal of Human-Computer Studies, 40*, 153–184.

Licklider, J. C. R. (1960). Man-computer symbiosis. *Institute of Radio Engineers Transactions on Human Factors in Electronics, HFE-1*, 4–11.

Mackworth, N. H. (1948). The breakdown of vigilance during prolonged visual search. *Quarterly Journal of Experimental Psychology, 1*, 6–21.

March, J. G., & Simon, H. A. (1958). *Organizations*. New York: Wiley.

Maurino, D. E. (2000). Human factors and aviation safety: What the industry has, what the industry needs. *Ergonomics, 43*, 952–959.

Mazlish, B. (1967). The fourth discontinuity. *Technology and Culture 8*(1).

McRuer, D. T., & Jex, H. R. (1967). A review of quasi-linear pilot models. *IEEE Transactions on Human Factors in Electronics, HFE-4(3)*, 231–249.

Meister, D. (1964). Methods of predicting human reliability in man-machine systems. *Human Factors, 6*, 621–646.

Mendel, M., & Sheridan, T. B. (1989). Filtering information from human experts. *IEEE Transactions on Systems, Man and Cybernetics, 36*(1), 6–16.

Meyer, J. (2000). Adaptive changes in the reliance on hazard warnings. *In Proceedings Human interaction with complex systems 2000* (pp. 39–42). Urbana-Champaign, IL: Beckman Institute.

Miller, C. A. (2000, March). *Rules of etiquette, or how a mannerly AUI should comport itself to gain social acceptance and be perceived as gracious and well behaved in polite society*. Unpublished working notes of AAAI Spring Symposium on Adaptive User Interfaces.

Miller, G. A. (1956). The magical number 7, plus or minus 2: Some limits on our capacity for processing information. *Psychological Review, 63*, 81–97.

Mitchell, C. M. (1987). GT-MSOCC: A research domain for modeling human-computer interactions and aiding decision-making in supervisory control systems. *IEEE Transactions on Systems, Man and Cybernetics, SMC-17*(4), 553–570.

Mitchell, C. M., & Forren, M. G. (1987). Effectiveness of multi-modal operator interfaces in supervisory control systems. *IEEE Transactions on Systems, Man and Cybernetics, SMC-17*(4), 594–607.

Mitchell, C. M., & Saisi, D. S. (1987). Use of model-based qualitative icons and adaptive windows in workstations for supervisory control systems. *IEEE Transactions on Systems, Man and Cybernetics, SMC-17*(4), 573–593.

Molloy, R., & Parasuraman, R. (1996). Monitoring an automated system for a single failure: Vigilance and task complexity effects. *Human Factors, 38*, 311–322.

Moray, N. (Ed.). (1979). *Mental workload: Its theory and measurement*. New York: Plenum.

Moray, N. (1986). Monitoring behavior and supervisory control. In K. R. Boff, L. Kaufman, & J. P. Thomas (Eds.), *Handbook of perception and human performance* (Vol. II, pp. 40-1–40.51). New York: Wiley.

Moray, N. (1997a). Human factors in process control. In G. Salvendy (Ed.), *Handbook of human factors and ergonomics* (2nd ed., pp. 1944–1971). New York: Wiley.

Moray, N. (1997b). Models of models of mental models. In T. B. Sheridan & T. van Lunteren (Eds.), *Perspectives on the human controller* (pp. 271–285). Mahwah, NJ: Erlbaum.

Moray, N., Inagaki, T., & Itoh, M. (2000). Adaptive automation, trust and self-confidence in fault management of time-critical tasks. *Journal of Experimental Psychology: Applied, 6*, 44–58.

Mouloua, M., & Koonce, J. M. (1997). *Human-automation interaction*. Mahwah, NJ: Erlbaum.

Muir, B. M. (1988). Trust between humans and machines, and the design of decision aids. In E. Hollnagel, G. Mancini, & D. D. Woods (Eds.), *Cognitive engineering in complex dynamic worlds* (pp. 71–84). London: Academic.

Muir, B. M., & Moray, N. P. (1996). Trust in automation. Part II. Experimental studies of trust and human intervention in a process control simulation. *Ergonomics, 39*, 429–461.

Mumaw, R. J., Roth, E. M., Vicente, K. J., & Burns, C. M. (2000). There is more to monitoring a nuclear power plant than meets the eye. *Human Factors, 42*, 36–55.

Murray, J., & Liu, Y. (1997a). Hortatory operations in cognitive engineering. International *Journal of Cognitive Ergonomics, 1*(2), 107–118.

Murray, J., & Liu, Y. (1997b). Hortatory operations in highway traffic management. *IEEE Transactions on Systems, Man and Cybernetics, Part A: Systems and Humans, 27*, 340–350.

Negroponte, N. (1995). *Being digital*. New York: Random House.

Neisser, U. (1976). *Cognition and reality: Principles and implications of cognitive psychology*. San Francisco: Freedman.

Newall, A. (1990). *Unified theories of cognition*. Cambridge, MA: Harvard University Press.

Norman, D. A. (1981). Categorization of action slips. *Psychological Review, 88*, 1–15.

O'Donnell, R. D., & Eggemeier, F. T. (1986). Workload assessment methodology. In K. Boff, L. Kaufman, & J. P. Thomas (Eds.), *Handbook of perception and human performance: Vol. 1. Sensory processes and perception* (pp. 42.1–42.49). New York: Wiley.

Oltmans, W. L. (Ed.). (1974). *On growth*. New York: Putnam.

Ottensmeyer, M. P., Hu, J., Thompson, J. M., Ren, J., & Sheridan, T. B. (2000). Investigation into performance of minimally invasive surgery with feedback time delays. *Presence: Teleoperators and Virtual Environments, 9*(4), 369–382.

Parasuraman, R., Hancock, P. A., & Olofinboba, O. (1997). Alarm effectiveness in driver-centered collision warning systems. *Ergonomics, 39*, 390–399.

Parasuraman, R., Masalonis, A. J., & Hancock, P. A. (2000). Fuzzy signal detection theory: Basic postulates and formulas for analyzing human and machine performance. *Human Factors, 42*, 636–659.

Parasuraman, R., & Mouloua, M. (Eds.). (1996). *Automation and human performance: Theories and applications*. Mahwah, NJ: Erlbaum.

Parasuraman, R., Mouloua, M., & Molloy, R. T. (1996). Effects of adaptive task allocation on monitoring of automated systems. *Human Factors, 38*, 665–679.

Parasuraman, R., & Riley, V. A. (1997). Humans and automation: Use, misuse, disuse, and abuse. *Human Factors, 39*, 230–253.

Parasuraman, R., Sheridan, T. B., & Wickens, C. D. (2000). A model for types and levels of human interaction with automation. *IEEE Transactions on Systems, Man and Cybernetics, SMC-30*(3), 286–297.

Park, J. H. (1991). *Supervisory control of robot manipulators for gross motions*. Ph.D. thesis, Massachusetts Institute of Technology, Cambridge, MA.

Patrick, N. J. M., & Sheridan, T. B. (1998, October). Modeling decision-making for vertical navigation of long-haul aircraft. In *Proceedings of the IEEE International Symposium on Cybernetics and Society*. La Jolla, CA.

Perrow, C. P. (1984). *Normal accidents*. New York: Basic Books.

Pew, R., & Mavor, A. (1998). *Modeling human and organizational behavior*. Washington, DC: National Academy Press.

Price, H. E. (1985). The allocation of functions in systems. *Human Factors, 27*, 33–45.

Price, H. E. (1990). Conceptual system design and the human role. In H. Booher (Ed.), *MANPRINT: An approach to systems integration* (pp. 161–204). New York: Van Nostrand.

Pritchett, A. (1999). Plot performance at collision avoidance during closely spaced parallel approaches. *Air Traffic Control Quarterly, 7*(1), 47–75.

Raiffa, H., & Schlaifer, R. (1961). *Applied statistical decision theory.* Cambridge, MA: Harvard University Graduate School of Business Administration.

Rasmussen, J. (1976). Outlines of a hybrid model of the process plant operator. In T. Sheridan & G. Johannsen (Eds.), *Monitoring behavior and supervisory control* (371–384). New York: Plenum.

Rasmussen, J. (1986). *Information processing in human-machine interaction.* Amsterdam: North-Holland.

Rasmussen, J. (2000). Human factors in a dynamic information society. *Ergonomics, 43,* 869–879.

Rasmussen, J., Pejtersen, A.M., & Goodstein, L. P. (1994). *Cognitive systems engineering.* New York: Wiley.

Rasmussen, J., & Vicente, K. (1992). Ecological interfaces: A technological imperative in high tech systems? *International Journal of Human-Computer Interaction, 2,* 93–111.

Rawls, J. (1999). *A theory of justice* (rev. ed.). New York: Belknap.

Reason, J. (1990). *Human error.* Cambridge, England: Cambridge University Press.

Reason, J. (1994). Foreword. In M. S. Bogner (Ed.), *Human error in medicine* (pp. vii–xv). Mahwah, NJ: Erlbaum.

Reason, J. (1997). *Managing risks of organizational accidents.* London: Ashgate Gower.

Reid, G. B., Shingledecker, C. A., & Eggemeier, F. T. (1981). Application of conjoint measurement to workload scale development. In *Proceedings of the Human Factors Society 25th Annual Meeting* (pp. 522–526). Santa Monica, CA: Human Factors and Ergonomics Society.

Repperger, D. W., Phillips, C. A., & Chelette, T. L. (1995). A study of spatially induced "virtual force" with an information theoretic investigation of human performance. *IEEE Transactions on Systems, Man and Cybernetics, 25*(1), 1392–1404.

Riley, V. (1994). A theory of operator reliance on automation. In M. Mouloua & R. Parasuraman (Eds.), *Human performance in automated systems: Recent research and trends* (pp. 8–14). Mahwah, NJ: Erlbaum.

Roseborough, J. (1988). *Aiding human operators with state estimates.* Unpublished Ph.D. thesis, Massachusetts Institute of Technology, Cambridge, MA.

Rouse, W. B. (1980). *System engineering models of human-machine interaction.* Amsterdam: North-Holland.

Rouse, W. B. (1988). Adaptive aiding for human/computer control. *Human Factors, 30,* 431–438.

Rouse, W. B., & Hammer, J. M. (1991). Assessing the impact of modeling limits on intelligent systems. *IEEE Transactions on Systems, Man and Cybernetics, SMC-21*(6), 1549–1559.

Rouse, W. B., Hammer, J. M., & Lewis, C. M. (1989). On capturing humans' skills and knowledge: Algorithmic approaches to model identification. *IEEE Transactions on Systems, Man and Cybernetics, SMC-19*(3), 558–573.

Rouse, W. B., & Hunt, R. M. (1984). Human problem solving in fault detection tasks. In W. B. Rouse (Ed.), *Advances in man-machine systems research* (Vol. 1, pp. 195–222). Greenwich, CT: JAI Press.

Rouse, W. B., & Morris, N. M. (1986). Looking into the black box: Prospects and limits in the search for mental models. *Psychological Bulletin, 100*(3), 349–363.

Ruffell-Smith, H. P. A. (1979). *Simulator study of the interaction of pilot workload with error, vigilance and decisions* (NASA TM-78482). Moffett Field, CA: NASA Ames Research Center.

Sage, A. P., & Rouse, W. B. (Eds.). (1999). *Handbook of systems engineering and management.* New York: Wiley.

Salvendy, G. (Ed.). (1997). *Handbook of human factors and ergonomics* (2nd ed.). New York: Wiley.

Sanders, M. S., & McCormick, E. J. (1987). *Human factors in engineering and design* (6th ed.). New York: McGraw-Hill.

Sarter, N. B., & Amalberti, R. (Eds.). (2000). *Cognitive engineering in the aviation domain.* Mahwah, NJ: Erlbaum.

Sarter, N. B., & Woods, D. D. (1997). Team play with a powerful and independent agent: Operational experiences and automation surprises on the Airbus A-320. *Human Factors, 39,* 553–569.

Sarter, N. B., Woods, D. D., & Billings, C. E. (1997). Automation surprises. In G. Salvendy (Ed.), *Handbook of human factors and ergonomics* (2nd ed., pp. 1926–1943). New York: Wiley.

Scerbo, M. W., & Mouloua, M. (Eds.). *Automation technology and human performance: Current research and trends.* Mahwah, NJ: Erlbaum.

Senders, J. W., Elkind, J. I., Grignetti, M. C., & Smallwood, R. P. (1964). *An investigation of the visual sampling behavior of human observers* (NASA-CR-434). Cambridge, MA: Bolt, Beranek and Newman.

Senders, J. W., & Moray, N. P. (1991). *Human error: Cause, prediction and reduction.* Mahwah, NJ: Erlbaum.

Shannon, C. E. (1949). Communication in the presence of noise. *Proceedings of the Institute of Radio Engineers, 37,* 10–22.

Shell, R. L., & Hall, E. L. (2000). *Handbook of industrial automation.* New York: Marcel Dekker.

Shepard, R. N., Romney, A. K., & Nerlove, S. B. (1972). *Multidimensional scaling.* New York: Seminar Press.

Sheridan, T. B. (1970). On how often the supervisor should sample. *IEEE Transactions on Systems Science and Cybernetics, SSC-6,* 140–145.

Sheridan, T. B. (1975). Community dialog technology. *Proceedings of the IEEE, 63*(3), 463–475.

Sheridan, T. B. (1980). Computer control and human alienation. *MIT Technology Review, 83*(1), 60–73.

Sheridan, T. B. (1983). Measuring, modeling and augmenting reliability of man-machine systems. *Automatica, 19.*

Sheridan, T. B. (1988). Trustworthiness of command and control systems. In *Proceedings of the International Federation of Automatic Control Conference on Man-Machine Systems* (pp. 427–431). Elmsford, NY: Pergamon.

Sheridan, T. B. (1992a). Musings on telepresence and virtual presence. *Presence, 1,* 120–125.

Sheridan, T. B. (1992b). *Telerobotics, automation and human supervisory control.* Cambridge: MIT Press.

Sheridan, T. B. (1993). Space teleoperation through time delay: Review and prognosis. *IEEE Transactions on Robotics and Automation, 9*(5), 592–606.

Sheridan, T. B. (1995). Reflections on information and information value. *IEEE Transactions on Systems, Man and Cybernetics, 25*(1), 194–196.

Sheridan, T. B. (1997). Supervisory control. In G. Salvendy (Ed.), *Handbook of human factors and ergonomics* (2nd ed., pp. 1295–1327). New York: Wiley.

Sheridan, T. B. (1998). Rumination on automation. In *Proceedings of the International Federation on Automatic Control Symposium on Man-Machine Systems*. Elmsford, NY: Pergamon.

Sheridan, T. B. (1999). Descartes, Heidegger, Gibson and God: Toward an eclectic ontology of presence. *Presence, 8*(5), 549–557.

Sheridan, T. B. (2000). Function allocation: Algorithm, alchemy, or apostasy? *International Journal of Human-Computer Studies, 52*, 203–216.

Sheridan, T. B., & Ferrell, W. R. (1974). *Man-machine systems*. Cambridge: MIT Press.

Sheridan, T. B., & Parasuraman, R. (2000). Human vs. automation in responding to failures: An expected value analysis. *Human Factors, 42*, 403–407.

Sheridan, T. B., & Simpson, R. W. (1979, January). *Toward the definition and measurement of the mental workload of transport pilots* (Report R79-4). Cambridge: Massachusetts Institute of Technology, Flight Technology Laboratory.

Sheridan, T. B., & Verplank, W. L. (1978). *Human and computer control of undersea teleoperators* (Man-Machine Systems Laboratory Report). Cambridge: MIT.

Slovic, P., Fischhoff, B., & Lichtenstein, S. (1977). Behavioral decision theory. *Annual Review of Psychology, 28*, 1–39.

Smith, K., & Hancock, P. A. (1995). Situation awareness is adaptive, externally directed consciousness. *Human Factors, 37*, 137–148.

Smith, P. J., McCoy, C. E., & Layton, C. (1997). Brittleness in the design of cooperative problem-solving systems: The effects on user performance. *IEEE Transactions on Systems, Man and Cybernetics, Part A: Systems and Humans, 27*, 360–371.

Snow, C. P. (1959). *Two cultures and the scientific revolution*. New York: Cambridge University Press.

Stauch, B. (2000, October). *Automation, workload and situation awareness*. Paper presented at the Conference on Human Performance, Situation Awareness and Automation, Savannah, GA.

Stewart, T. (2000). Ergonomics user interfaces standards: Are they more trouble than they are worth? *Ergonomics, 43*, 1030–1044.

Suchman, L.A. (1987). *Plans and situated actions: The problem of human-machine communication*. Cambridge, England: Cambridge University Press

Swain, A. D., & Guttman, H. E. (1983). *Handbook of human reliability analysis with emphasis on nuclear power plant applications* (Sandia National Labs/NUREG CR-1278). Washington, DC: U.S. Nuclear Regulatory Commission.

Thompson, J. M., Ottensmeyer, M. P., & Sheridan, T. B. (1999, October). Human factors in tele-inspection and tele-surgery: Cooperative manipulation under asynchronous video and control feedback. In *Proceedings of the Medical Image Computing and Computer-Assisted Intervention 1st International Conference* (pp. 368–377.)

Tufte, E. R. (1983). *The visual display of quantitative information*. Cheshire, CT: Graphics Press.

Tufte, E. R. (1997). *Visual explanations*. Cheshire, CT: Graphics Press.

Tulga, M. K., & Sheridan, T. B. (1980). Dynamic decisions and workload in multi-task supervisory control. *IEEE Transactions on Systems, Man and Cybernetics, SMC-10*(5), 217–231.

Turkle, S. (1995). *Life on the screen: Identity in the age of the Internet*. New York: Simon and Schuster.

Tversky, A., & Kahneman, D. (1973). Availability, a heuristic for judging frequency and probability. *Cognitive Psychology, 5*, 207–232.

Tversky, A., & Kahneman, D. (1974). Judgment under uncertainty: Heuristics and biases. *Science, 185*, 1124–1131.

Tversky, A., & Kahneman, D. (1980). *Causal schemes in judgments under uncertainty.* New York: Cambridge University Press.

Tversky, A., & Kahneman, D. (1981). The framing of decisions and the psychology of choice. *Science, 211,* 453–458.

Tversky, A., & Kahneman, D. (1992). Advances in prospect theory: Cumulative representation of uncertainty. *Journal of Risk and Uncertainty, 5,* 297–323.

U.S. Nuclear Regulatory Commission. (1981). *NUREG-0700. Guidelines for control room design reviews.* Washington, DC: Author.

Van Cott, H. P., & Kinkade, R. G. (1972). *Human engineering guide to equipment design.* Washington, DC: American Institutes for Research.

Vanderhaegen, F., Crevits, I., Debernard, S., & Millot, P. (1994). Human-machine cooperation: Toward activity regulation assistance for different air traffic control levels. *International Journal of Human-Machine Interaction, 6,* 65–104.

Vertut, J., & Coiffet, P. (1986). *Robot technology: Vol. B. Teleoperation and robotics: Evolution and development.* Englewood Cliffs, NJ: Prentice-Hall.

Vicente, K. (1999). *Cognitive work analysis.* Mahwah, NJ: Erlbaum.

Vicente, K., Moray, N., Lee, J. D., Rasmussen, J., Jones, B. G., Brock, R., & Djemil, T. (1996). Evaluation of a Rankine cycle display for nuclear power plant monitoring and diagnosis. *Human Factors, 38,* 506–521.

Vicente, K., & Rasmussen, J. (1992). Ecological interface design: Theoretical foundations. *IEEE Transactions on Systems, Man and Cybernetics, SMC-22,* 589–606.

Von Neumann, J., & Morganstern, O. (1944). *Theory of games and economic behavior.* New York: Princeton University Press.

Weber, E. (1994). From subjective probabilities to decision weights: The effect of asymmetric loss functions on the evaluation of uncertain events. *Psychological Bulletin, 115,* 228–242.

Wei, Z., Macwan, A. P., & Wierenga, P. A. (1998). A quantitative measure for degree of automation, and its relation to system performance. *Human Factors, 40,* 277–295.

Wichansky, A. M. (2000). Usability testing in 2000. *Ergonomics, 43,* 998–1006.

Wickens, C. D. (1994). Designing for situation awareness and trust in automation. In *Proceedings of International Federation of Automatic Control Conference on Integrated Systems Engineering.* Elmsford, NY: Pergamon.

Wickens, C. D., Gordon, S. E., & Liu, L. (1998). *An introduction to human factors engineering.* New York: Addison Wesley Longman.

Wickens, C. D., Mavor, A. S., Parasuraman, R., & McGee, J. P. (1998). *The future of air traffic control* (National Research Council report). Washington, DC: National Academy Press.

Wickens, C. D., Tsang, P., & Pierce, B. (1985). The dynamics of resource allocation. In W. B. Rouse (Ed.), *Advances in man-machine systems research* (Vol. 2, pp. 1–50). Greenwich, CT: JAI.

Wiener, E. L., & Curry, R. E. (1980). Flight deck automation, promises and problems. *Ergonomics, 23,* 995–1011.

Wiener, N. (1964). *God and golem, incorporated.* Cambridge: MIT Press.

Wierwille, W. W., & Casali, J. G. (1983). A validated rating scale for global mental workload measurement applications. In *Proceedings of Human Factors Society 27th Annual Meeting* (pp. 129–133). Santa Monica, CA: Human Factors and Ergonomics Society.

Wierwille, W. W., Casali, J. G., Connor, S. A., & Rahimi, M. (1985). Evaluation of the sensitivity and intrusion of mental workload estimation techniques. In W. B. Rouse (Ed.), *Advances in man-machine systems research* (Vol. 2, pp. 51–127). Greenwich, CT: JAI Press.

Winkler, R. L., & Murphy, A. H. (1973). Experiments in the laboratory and in the real world. *Organizational Behavior and Human Performance, 10*, 252–270.

Woods, D. D. (1996). Decomposing automation: Apparent simplicity, real complexity. In R. Parasuraman & M. Mouloua (Eds.), *Automation and human performance: Theory and application* (pp. 1–17). Mahwah, NJ: Erlbaum.

Woods, D. D., & Roth, E. M. (1995). Symbolic AI computer simulations as tools for investigating the dynamics of joint cognitive systems. In J.-M. Hoc, P. C. Cacciabue, & E. Hollnagel (Eds.), *Expertise and technology: Cognition and human-computer cooperation*. Mahwah, NJ: Erlbaum.

Xiao, Y., Milgram, P., & Doyle, J. D. (1997). Planning behavior and its functional role in interactions with complex systems. *IEEE Transactions on Systems, Man and Cybernetics, Part A: Systems and Humans, 27*, 313–324.

Yoerger, D. R. (1982). *Supervisory control of underwater telemanipulators: Design and experiment*. Unpublished Ph.D. thesis, Massachusetts Institute of Technology, Cambridge, MA.

Young, M. S., & Stanton, N. A. (1997). Automotive automation: Investigating the impact on drivers' mental workload. *International Journal of Cognitive Ergonomics, 1*(4), 325–336.

Zadeh, L. A. (1965). Fuzzy sets. *Information and Control, 8*, 338–353.

Zsambok, C. E., & Klein, G. A. (1997). *Naturalistic decision making*. Mahwah, NJ: Erlbaum.

Zuboff, S. (1988). *In the age of the smart machine: The future of work and power*. New York: Basic Books.

Author Index

Adams, M. J., 17
Adelman, L., 105
Ainsworth, L. K., 56
Amalberti, R., x, 15, 18
Arkes, H. R., 74
Arrow, K. J., 142
Askey, S. Y., 29
Askren, W. B., 94

Bainbridge, L., 152, 158, 160, 172
Barfield, W., 23, 37
Bar Hillel, M., 73
Baron, S., 71, 72, 158
Billings, C. E., 15, 18, 85, 155
Birmingham, H. P., 59
Bisantz, A. M., 196
Bisseret, A., 60
Blackman, H. S., 41
Blumer, C, 74
Boff, K., 88
Booher, H. R., 7
Boy, G. A., 55
Bresnick, T. A., 105
Brock, R., 100
Brooks, T. L., 35, 121
Brunswik, E., 195
Buckner, D. N., 79
Bunzo, M., 158
Burdea, G. C., 35
Burns, C. M., 105

Cacciabue, P. C., 89
Campbell, D. T., 136, 138
Carbonell, J., 192
Casali, J. G., 79, 81
Chapanis, A., 98
Charny, L., 112, 120
Chelette, T. L., 104
Chen, S., 25
Chinnis, J. O., 105
Christoffersen, K., 99

Cohen, M., 105
Coiffet, P., 33
Connor, S. A., 81
Cook, R. I., 43
Cook, T. D., 136, 138
Cooksey, R. W., 196
Corker, K., 158
Corrigan, J. M., 83
Craik, K. J. W., 58
Crevits, I., 60
Curry, R. E., 55, 123, 172

Dalkey, N., 141
Das, H., 35
Debernard, S., 60
Degani, A., 18, 94
Dekker, S., 18
Desmond, P. A., 82
DeSola Pool, I., 179
Dingus, T. A., 23
Djemil, T., 100
Donaldson, M. S., 83
Dowell, J., 56
Doyle, J. D., 43
Draper, J. V, 36
Durlach, N. I., 37, 104

Eby, D. W., 24
Edwards, W., 73
Eggemeier, F. T., 79
Einhorn, H. J., 73, 74
Einstein, A., 59
Elkind, J. I., 192
Ellis, S. R., 91
Embry, D. E., 93
Endsley, M., 75, 76, 122
Engelberger, J., 175
Evans, J. B. T., 74

Ferrell, W. R., 36, 199
Fischhoff, B., 73, 173

Fisk, A. D., 196
Fitts, P. M., 58, 70, 89, 155
Flach, J. M., 75, 156
Forren, M. G., 104
Fromm, E., 174
Fuld, R. B., 60
Furness, T. A., 37

Gai, E. G., 123
Garner, W., 98
Gay, P., 196
Gertman, D., I. 41
Gibson, J. J., 99, 128, 156, 159
Gilmore, W. E., 41
Goodstein, L. P., 56
Gordon, S. E., 7
Green, D. M., 210
Greenberg, S., 48
Grice, H. P., 112, 113
Grignetti, M. C., 192
Guttman, H. E., 91

Hall, E. L., 11, 67
Hall, R. E., 92
Hammer, J. M., 155, 158
Hammond, K. R., 195
Hancock, P. A., 64, 75, 82, 211, 217–219
Hansman, R. J., 102
Hardin, G., 179
Hart, S. G., 77
Heidegger, M., 99, 128, 159
Helander, M. G., 67
Helmer, O., 141
Helmreich, R. L., 90
Hick, W. E., 70
Hoc, J.-M., 59
Hogarth, R. M., 73, 74
Hollnagel, E., 18, 83
Howard, R. A., 199
Hu, J., 45
Hunt, R. M., 86
Hunter, C. N., 99
Hursch, C J., 195
Hursch, J. L., 195

Inagaki, T., 64, 99
Infeld, L., 59
Isermann, R., 229
Itoh, M., 64

Jasek, C. A., 58, 108
Jex, H. R., 71, 72, 158
Johnson-Laird, P. N., 158
Jones, B. G., 100
Jones, P. M., 59, 108
Jordan, N., 58, 59, 60

Kaber, D. B., 36, 64, 75, 172
Kahneman, D., 73, 74, 158
Kalman, R. E., 123, 136
Kantowitz, B., 60
Kappel, G., 136, 138
Keeney, R. L., 66, 203
Kinkade, R. G., 4, 67
Kirlik, A., 94, 128, 196
Kirwan, B., 56
Klein, G. A., 97, 99, 156, 158
Kleinman, D. L., 71, 72, 158
Kohn L. T., 83
Komoda, N., 25
Koonce, J. M., x
Kostyniuk, L. P., 24
Kozinsky, E., 70
Kreith, F., 67
Kuchar, J. K. 102, 106
Kusunoki, H., 25

Landauer, T. K., 67
Lanzilotta, E. J., 93
Laskey, K. B., 105
Layton, C., 148
Lee, J. D., 77, 100, 148
Levison, W. H., 71, 72, 158
Lewis, C. M., 158
Lichtenstein, S 73, 173
Licklider, J. C. R., 155
Lincoln, J., 88
Liu, L., 7
Liu, Y., 26, 128
Long, J., 56

Mackworth, N. H., 107
Macwan, A. P., 63
March, J. G., 120
Masalonis, A. J., 217–219
Maurino, D. E., 90
Mavor, A. S., 14, 20, 37, 63, 72, 104, 155
Mazlish, B., 175
McCormick, E. J., 7

McCoy, C. E., 148
McGee, J. P., 14, 20, 63, 155
McGrath, J. J., 79
McRuer, D. T., 71, 72, 158
Meister, D., 91
Mendel, M., 220
Merritt, A. C., 90
Meyer, J., 124
Milgram, P., 43
Miller, C. A., 112, 113
Miller, G. A., 70
Millot, P., 60
Mitchell, C. M., 103, 104, 108
Molloy, R. T., 64, 107
Monta, K., 156
Moray, N. P., 64, 76, 83, 77, 100, 148, 158, 192
Morgan, C., 98
Morganstern, O., 201, 204, 233
Morris, N. M., 158
Mouloua, M., x, 64
Muir, B. M., 77
Mumaw, R. J., 105
Murphy, A. H., 73
Murray, J., 26, 128

Negroponte, N., 51
Neisser, U., 75
Nerlove, S. B., 141
Newall, A., 158
Norman, D. A., 84

O'Donnell, R. D., 79
Olofinboba, O., 211
Oltmans, W. L., 179
Ottensmeyer, M. P., 45, 101

Pack, R., 70
Parasuraman, R., x, 14, 20, 55, 63, 64, 77, 107, 154, 155, 209, 211, 217–219
Park, J. H., 35
Patrick, N. J. M., 22
Pejtersen, A. M., 56
Perrow, C. P., 83, 85
Pew, R. W., 17, 72
Phillips, C. A., 104
Phipps, D. A., 196
Pierce, B, 122
Pisanich, G., 158
Prabhu, P. V., 67

Price, H. E., 59, 60
Pritchett, A., 125

Rahimi, M., 81
Raiffa, H., 66, 199, 203
Rasmussen, J., 7, 56, 60, 99, 100, 145, 156
Rawls, J., 176
Reason, J., 83, 84, 85, 86
Regulinski, T. L., 94
Reid, G. B., 79
Ren, J., 45
Repperger, D. W., 104
Riley, J. M., 36, 64, 172
Riley, V. A., 77, 154, 211
Romney, A. K., 141
Roseborough, J., 149
Roth, E. M., 56, 105
Rouse, W. B., 64, 86, 155, 158, 182
Ruffell-Smith, H. P. A., 83

Sage, A. P., 182
Saisi, D. S., 103
Salvendy, G., 67
Samanta, P. K., 92
Sanders, M. S., 7
Sarter, N. B., x, 15, 18, 85
Scallen, S. F., 64
Schlaifer, R., 199
Seminara, J., 70
Senders, J. W., 83, 192
Shafto, M., 94
Shannon, C. E., 70, 196, 197, 200
Shell, R. L., 11, 67
Shepard, R. N., 141
Sheridan, T. B., 20, 22, 25, 33, 36, 45, 55, 60, 62, 66, 70, 76, 77, 83, 93, 94, 101, 104, 108, 112, 115, 141, 156, 164, 170, 179, 192, 199, 200, 209, 220
Shingledecker, C. A., 79
Simon, H. A., 120
Simpson, R. W., 79
Slovic, P., 73, 158, 173
Smallwood, R. P., 192
Smith, K., 75
Smith, P. J., 148
Snow, C. P., 186
Sorkin, R., 60
Stanton, N. A., 82
Stauch, B., 14

Stewart, T., 68
Strauch, B., 75
Suchman, L. A., 156
Swain, A. D., 91
Swets, J. A., 210
Swoboda, A. L., 92

Tanabe, F., 156
Taylor, F. V., 59
Tenney, Y. J., 17
Thompson, J. M., 45, 101
Tsang, P., 122
Tufte, E. R., 95, 102
Tulga, M. K., 122
Turkle, S., 172
Tversky, A., 73, 74, 158

U.S. Nuclear Regulatory Commission, 58, 67
Usher, J. M., 36

Van Cott, H. P., 4, 67
Vanderhaegen, F., 60
Verplank, W. L., 62
Vertut, J., 33

Vicente, K., 56, 99, 100, 105, 156
Von Neumann, J., 201, 204, 233
Vruels, D., 70

Walker, N., 196
Weber, E., 74
Wei, Z., 63
Wichansky, A. M., 141
Wickens, C. D., 7, 14, 20, 63, 55, 77, 122, 155
Wiener, E. L., 18, 55, 172
Wiener, N., 173, 181
Wierenga, P. A., 63
Wierwille, W. W., 79, 81
Winkler, R. L., 73
Woods, D. D., 18, 43, 55, 56, 83, 85

Xiao, Y., 43

Yoerger, D. R., 35, 121
Young, M. S., 82

Zadeh, L. A., 205
Zsambok , C. E., 97, 99, 156
Zuboff, S., 170

Subject Index

Absolute judgment, 70
Accident data banks, 55
Advocacy, 168
Aircraft, 14
    piloting, 14
Air traffic control, 18
Alarm, 104
    complexity, 105
    inhibition, 105
    nuisance, 106
Alienation, 170, 175
Allocation stages, 61
All-or-none fallacy, 150
Analogic display, 106
American National Standards Institute
    (ANSI), 67
Apollo program, 154
Authority in human-machine system, 152
Automation, 9, 14, 150
    adaptive, 64
    advantages of, 163
    level of, 62
    on the person, 51
Automobile, 23
    navigation, 23
    traffic management, 26

Bayes' theorem, 190
Biases of decision makers, 72, 166
Bode plot, 226
Brunswik lens, 195

Calibration of decision maker, 219
Capture error, 85
Cathedral building, 165
Characteristic equation, 226
Chart junk, 102
Cognition, 157
Command and control, 46
Combinatorial analysis, 90
Command language, 35

Commission error, 84
Compensatory display, 71
Complacency, 147
Concurrent engineering, 184
Control, 220
    gain, 224
    stability, 225
Cost trade-offs, 184
Counterbalancing, 138
Crisp rules, 204

Data-link, 21
Decision, 72
    criterion, 208
    under certainty, 203
    under uncertainty, 207
Decision aid, 95, 148
Delphi method, 141
Design, 54, 187
Deskilling, 172
Desocialization, 172
Display
    adaptive format, 98
    design, 95
    integration, 97
    keyhole problem, 95
Distributive justice, 176
Dominating strategy, 212
Draper, Charles Stark, 154

Ecoexploitation, 179
Ecological display, 56, 99, 156
Education, 182, 186
Embodied intelligence, 165
Enslavement, 175
Ergonomics, 2
Error, 82
    causes, 84
    cost, 192
    definition, 82
    latent property, 86

remedies, 87
taxonomies, 84
Estimation theory, 135
Etiquette, 112
Evaluation, 130, 140
Event tree, 91
Experimental design, 136

Fatigue, 82
Fault tree, 91
Feedback control, 220
Fitts MABA-MABA list, 58
Fitts' law, 70
Flexible manufacturing, 41
Flight director, 154
Flight management system (FMS), 14
Focus group, 141
Framing, 74
Free flight, 21
Free will, 159
Function allocation, 57
Fuzzy control, 227
Fuzzy logic, 161
Fuzzy rules, 205
Fuzzy signal detection, 217

Gain-phase plot, 226
Game theory, 212
Guidelines, 67

Handbooks, 67
Haptics, 38, 103
Head-related transfer function, 38
Hick's law, 70
Hospital systems, 43
Human factors, 2, 6, 41, 54, 67, 87, 169, 184
Human Factors and Ergonomics Society, 6
Human factors professional, 54
Human irrationality, 165
Human performance, 69
bandwidth, 147
Human-automation system, 2, 53, 66, 115, 130, 138, 151, 157, 161, 182, 192
Human-centered automation, 155
Human-interactive computer, 16
Human-in-the-loop simulator, 131

Icon, 103
Institute of Electrical and Electronics Engineers (IEEE), 67
Illiteracy, technological, 173
Implantable sensors, 52
Impossibility theorem, 142
Information, 194
quality, 194
Shannon, 197
value of, 199
Intervention, 124
Intimidation, 172
International Standards Organization (ISO), 67

Jack, 185
Jastrobowski, Wojciech, 2
Joint cognition, 56

Kalman estimator, 136

LaPlace transform, 222
Large system organization, 65
Learning, 126
Life cycle, 182
Limits of modeling, 157
Loop equations, 223

Manufacturing cell, 41
Markov network, 93
Mean time between failure, 94
Medical monitoring, 43
Membership function, 205
Memory, short term, 70
Mental model, 159
Mental workload, 77
physical workload, 77
subjective rating, 79
task load, 78
transients, 79
Menu, 103
Meta-knowledge, 113
Minimally invasive surgery, 45
Minimax strategy, 212
Mixed initiatives, 145
Mode error, 89
Model, 186, 189
Modern control, 227
Monitoring, 107, 121
Monkey's Paw, 173

Monte Carlo simulation, 93
Multiattribute utility, 202
Multidimensional scaling, 141
Mystification, 173

National Advanced Driving Simulator
    (NADS), 133
Naturalistic display, 99
Nuclear power plant, 38
Null hypothesis, 192

Objective function, 200
Office systems, 47
Omission error, 84
On-off control, 221
Optimal control, 71, 227
Optimization, 66, 130

Pareto frontier, 151, 203
Part-task simulator, 132
Peer influence, 169
Performance-shaping factors, 92
Planning, 119
Poles and zeroes, 226
Population stereotype, 101
Predictor display, 108
Prisoner's dilemma, 213
Probability, 189
Process control, 38
Procrustean bed, 160
Productive orientation, 174
Productivity, 177
Proportional-plus-derivative control, 59
Prospect theory, 74
Psychophysical scales, 194

Quasi-separability, 203

Recognition-primed decision, 99
Regret, 74
Relative operating characteristic (ROC),
    216
Reliability analysis, 89
Remote control, 164, 178
Representativeness, 74
Response time, 69
Responsibility, abandonment of, 174
Risk homeostasis, 86
Roseborough dimemma, 149

Safety, 169
    state, 93
Sampling, 191
Satisficing, 110
Secondary task, 79
Self-pacing, 72
Sensitivity analysis, 92
Ship, 31
    containerization, 31
    crew size, 31
    predictor display, 32
Signal detection theory, 213
Simple crossover model, 71
Simulation, 130
Simulator fidelity, 131
Situation awareness, 75
    assessment technique, 76
Skill, rule, and knowledge, 7
Skinner, B. F., 48
Smart checklist, 107
Smart home, 49
Social choice, 142
Social issues, 163
Soldiers, spies, and saboteurs, 180
Sorcery, 167
Spacecraft, 32
Stability, 225
Standards, 67, 167
State variable, 220
Stimulus-response compatibility, 100
Stress, 82, 86
Subsumption architecture, 147
Supervisory control, 43, 115
    definition of, 115
    historical trend, 116
    roles of human, 118
    useful range, 128
Symbolic display, 106
System complexity, 144
System management, 182, 185
Systems, Man and Cybenetics Society
    (IEEE), 6

Task analysis, 55
Teaching, 120
    machine, 48
Technophiles and technophobes, 176
Telegladiator, 180
Telegovernance, 178
Telemedicine, 45
Teleoperator, 32

Telepresence, 36
Telerobot, 32
Telesurgery, 45
Terrorism, 181
Testing, 130
Three Mile Island, 40
Time delay, 36
TRACON, 20
Trading and, sharing 63
Traffic alert and collision avoidance
    system (TCAS), 14
Train, 27
    accident, 154
    dispatching, 30
    displays in, 28
    driver's task, 27

Transfer function, 221
Trust, 18, 76, 148
Types of errors, 193

Underware, 51
Unemployment, 171
Usability testing, 50, 141
Utility function, 200

Virtual reality, 36, 101, 138
Voice recognition, 103
Von Neumann axiom, 201

Work dissatisfaction, 171
Worker control, loss of, 171

ALPHONSE CHAPANIS
Human Factors in Systems Engineering

YACOV Y. HAIMES
Risk Modeling, Assessment, and Management

DENNIS M. BUEDE
The Engineering Design of Systems: Models and Methods

ANDREW P. SAGE and JAMES E. ARMSTRONG, Jr.
Introduction to Systems Engineering

WILLIAM B. ROUSE
Essential Challenges of Strategic Management

YEFIM FASSER and DONALD BRETTNER
Management for Quality in High-Technology Enterprises

THOMAS B. SHERIDAN
Humans and Automation: System Design and Research Issues